Studies in Systems, Decision and Control

Volume 303

The series "Studies in Systems, Decision and Control" (SSDC) covers both new developments and advances, as well as the state of the art, in the various areas of broadly perceived systems, decision making and control–quickly, up to date and with a high quality. The intent is to cover the theory, applications, and perspectives on the state of the art and future developments relevant to systems, decision making, control, complex processes and related areas, as embedded in the fields of engineering, computer science, physics, economics, social and life sciences, as well as the paradigms and methodologies behind them. The series contains monographs, textbooks, lecture notes and edited volumes in systems, decision making and control spanning the areas of Cyber-Physical Systems, Autonomous Systems, Sensor Networks, Control Systems, Energy Systems, Automotive Systems, Biological Systems, Vehicular Networking and Connected Vehicles, Aerospace Systems, Automation, Manufacturing, Smart Grids, Nonlinear Systems, Power Systems, Robotics, Social Systems, Economic Systems and other. Of particular value to both the contributors and the readership are the short publication timeframe and the world-wide distribution and exposure which enable both a wide and rapid dissemination of research output.

** Indexing: The books of this series are submitted to ISI, SCOPUS, DBLP, Ulrichs, MathSciNet, Current Mathematical Publications, Mathematical Reviews, Zentralblatt Math: MetaPress and Springerlink.

More information about this series at http://www.springer.com/series/13304

Keyurkumar Patel · Axaykumar Mehta

Discrete-Time Sliding Mode Protocols for Discrete Multi-Agent System

Springer

Keyurkumar Patel
Institute of Infrastructure Technology
Research and Management (IITRAM)
Ahmedabad, Gujarat, India

Axaykumar Mehta
Institute of Infrastructure Technology
Research and Management (IITRAM)
Ahmedabad, Gujarat, India

ISSN 2198-4182 ISSN 2198-4190 (electronic)
Studies in Systems, Decision and Control
ISBN 978-981-15-6310-2 ISBN 978-981-15-6311-9 (eBook)
https://doi.org/10.1007/978-981-15-6311-9

This Springer imprint is published by the registered company Springer Nature Singapore Pte Ltd.
The registered company address is: 152 Beach Road, #21-01/04 Gateway East, Singapore 189721, Singapore

Keyurkumar Patel dedicated this book to his beloved parents, wife Dr. Hetal, and lovely daughter Urjita

and

Axaykumar Mehta dedicated this book to his mother Jayaben, wife Dr. Hema, and loving sons Siddharth, and Yatharth

Preface

The cooperative control of multiple systems popular as Multi-Agent System (MAS) has been widely researched area for the last few years due to its broad applications such as autonomous vehicles, military surveillance, smart/micro grid, vehicle traffic management, robotic teams, aerial robots, etc. Traditionally, a common centralized digital computer was employed to collect the information from the other system defined as an agent for controlling the multi-agent system. However, it increases the complexity due to the communication network and is also prone to failure due to system delay or corrupted data which deteriorates the performance of the MAS. Also, it requires more communication resources like large bandwidth and processing time. To overcome this kind of problems, the distributed cooperative control is proposed in the literature. The theme of distributed controllers lies in nature. The collective behaviors of animal groups in nature have shown that distributed decisions made by each individual for its own position, direction, and speed of motion can make the whole group behave like a single entity, which has its own rules of motion and decision-making power. The examples of such collective behavior can be seen in bird flocking, fireflies, schools of fish, and neuron firing. Inspired by these natural collective behaviors of animals in their groups, scientists and engineers got encouraged to develop distributed cooperative control for a group of systems by exchanging their information with neighbors.

The main idea in cooperative control of the MAS is to design the distributed controller for each agent by using its local neighboring information to arrive on a specific goal. In literature, many algorithms have been proposed for cooperative control, e.g., flocking, formation control, rendezvous, gossip, and consensus. However, among all the algorithms, the consensus algorithm is very popular in the fraternity. The research of the consensus algorithm is divided into two groups. The first group is working on the consensus without a leader or average consensus in which all the agents agree on a common point by interacting with each other. The other group focuses on leader-following consensus in which all the follower agents follow the leader. Many control methods have been developed and also successfully applied to the physical system for the leader-following consensus including H_∞, state feedback, Lyapunov approach, LMI, sliding mode control, etc… All the aforementioned algorithms are developed in the continuous-time domain. However, the discrete-time domain is quite advantageous as compared to the continuous-time domain for practical implementation using digital controllers. This is the reason why many scientists and researchers have opted for the leader-following consensus in the discrete-time domain. In literature, we may find the protocols in the discrete-time domain which lack the robustness and also achieve the consensus asymptotically.

This monograph presents novel discrete-time sliding mode protocols for the leader-following consensus of the discrete multi-agent system which achieves the consensus in finite time and also tackles the matched uncertainties. Further, the communication topology used to derive the protocols is represented by graph theory. Based on the communication graph topology, the protocols are divided into two groups, namely (i) fixed graph topology and (ii) switching graph topology. The fixed graph topology represents the fixed and static communication between the agents, while in the switching graph topology, the graph topology is dynamic and changes as per the number of agents participating during the consensus process.

First, a discrete-time sliding mode protocol is proposed for consensus of a linear discrete-time multi-agent system configured with a fixed, undirected graph topology as a global system having one leader and other agents as followers. The sliding surface for the global system is defined, and a discrete-time sliding mode protocol using well-known Gao's reaching law principle is derived which ensures consensus of followers with the leader in finite time. The sufficient condition for consensus, that is, the graph eigenvalues restricted into a unit circular region in the complex plane is derived. The robustness property of the proposed controller is also checked by applying a slowly varying disturbance. Finally, a numerical example is presented to endow the efficacy of the proposed methods. Time is an important property for the convergence of consensus. Fixed time steps for consensus are calculated using Gao's reaching law and Power rate reaching law-based protocols for leader-following consensus of DMAS comprising 2-DOF (degree of freedom) helicopter systems where the consensus of pitch angle and its velocity and yaw angle and its velocity are achieved. Both protocols were also verified for the consensus of DMAS with the application of a 2-DOF helicopter system with the switching graph communication topology in which agents may either enter or leave at a specific time. The protocol obtained using Gao's reaching law is suffering from chattering which deteriorates the leader-following consensus convergence speed of DMAS. Moreover, the protocol using Power rate reaching law improved the consensus convergence speed, but with the compromise of invariance property of SMC. To solve this problem, the higher order DSM protocols are designed using (i) reaching law approach and (ii) discrete super-twisting algorithm for the global consensus of DMAS in which chattering is reduced and invariance property is not lost. The proposed protocols ensure the consensus of the follower agents with the leader agent in finite time. The protocols are validated in simulation as well as experimentally on homogeneous DMAS comprising 2-DOF serial flexible joint robotic arms. In all the proposed protocols, it is mandatory to have continuous communication among all agents for the consensus. However, in many applications, the consensus is required only when a certain event occurs. Hence, to reduce the communication resources, a higher order discrete super-twisting sliding mode protocol for the leader-following consensus of discrete homogeneous multi-agent systems using an event-triggered approach is proposed. The protocol is designed based on the event-triggered strategies to achieve the consensus among the agents using intermittent communication with the neighboring agent. Each agent continuously observes its own state which is called a measurement error in order to check the triggering condition which needs to be satisfied for a particular agent control update. The proposed protocol is implemented on a discrete homogeneous multi-agent system setup comprising 2-DOF flexible joint robotic arms considering one robotic arm as a leader and others as followers. Further, the higher order discrete sliding mode control protocol using DSTA is proposed for the leader-following consensus of discrete heterogeneous linear multi-agent system. The proposed protocol ensures the consensus of heterogeneous follower agents with a leader agent in finite time. The algorithm is validated in simulation as well as experimentally on a 2-DOF flexible joint and 2-DOF flexible link robotic arms.

This monograph contributes to developing the robust and finite time or fixed time topological DSM protocols for the consensus of leader-following DMAS. These objectives are narrated below:

- Chapter 1 presents the brief introduction of the multi-agent system and the literature review of cooperative control algorithms for the consensus of homogeneous, heterogeneous MAS in continuous-time and discrete-time domains.
- In Chap. 2, the preliminaries of sliding mode control technique and the graph theory which are used for derivation of protocols in this monograph are discussed.
- In Chap. 3, the novel Discrete-time Sliding Mode (DSM) protocols are proposed for the consensus of linear DMAS by exchanging the information among themselves to achieve a common task. DMAS is represented using fixed, bidirectional, or undirected graph having one active leader and other as follower agents. A topological sliding surface is proposed for this study and also the number of steps required for the consensus is derived. The robustness of the proposed DSM protocols is also checked by applying matched disturbance to the DMAS.
- In Chap. 4, a design method is presented for the novel consensus protocols for the leader-following DMAS with switching graph topology. This proposed design scheme is applied to a 2-DOF helicopter system.
- In Chap. 5, the design methods of discrete higher order sliding mode protocols using reaching law, Discrete-time Super-Twisting Algorithm (DSTA) are proposed for the consensus of DMAS. The proposed method decreased the chattering effect and improves the consensus performance of DMAS. The proposed protocols are successfully validated on hardware setup available in the laboratory.

- In Chap. 6, the design method of an event-triggered higher order DSM protocol using DSTA for the leader-following consensus of DMAS is presented. The event-triggered scheme is very important to save the communication bandwidth, and hence more useful data can be sent through the channel. In this method, control is only updated when the triggering condition is satisfied and hence save the energy consumption. The proposed protocol is successfully applied to the hardware setup available in the laboratory.
- In Chap. 7, the design method of higher order DSM protocol using DSTA for the leader-following consensus of heterogeneous DMAS is proposed. In many real-time situations, the dynamics of each agent in a DMAS may not be necessarily same and hence it is interesting to study how the proposed method is useful to achieve the consensus for heterogeneous DMAS. The design method is successfully implemented on hardware setup consist of 2-DOF flexible joint and flexible link systems.

Ahmedabad, India
April 2020

Keyurkumar Patel
Axaykumar Mehta

Contents

About the Authors

Keyurkumar Patel obtained his M.Tech. degree from Nirma University, Ahmedabad, in 2014. Currently, he is pursuing his Ph.D. degree at Institute of Infrastructure Technology Research and Management (IITRAM), Ahmedabad in the broad area of control and robotics. Prior to joining IITRAM, he has worked as an Assistant Professor at the SVBIT, Gujarat Technological University, from 2014 to 2016. He has also worked as an Engineer at Instrumentation Ltd., a PSU under the Ministry of Heavy Industry. His research interests include co-operative control of multi-agent systems, nonlinear sliding mode controls and observers, sliding mode control applications in electrical engineering, and networked control of multi-agent systems. He has published several research papers in peer-reviewed international journals and conference proceedings, and is an active student member of the IEEE.

Axaykumar Mehta received the Ph.D. degree in 2009 from Indian Institute of Technology Bombay. He is currently an Associate Professor in Electrical Engineering at the Institute of Infrastructure Technology Research and Management, Ahmedabad. Prior to that, he was Director of Gujarat Power Engineering and Research Institute, Mehsana, Gujarat. He has published more than 90 research articles and book chapters in reputed journals, conference proceedings, and books. He has authored 3 monographs on sliding mode control and edited 3 conference proceedings with Springer Nature Singapore. He has also published 5 patents at the Indian Patent Office Mumbai. He is an advisor for the Design Lab project and Robotics museum at Gujarat Council on Science and Technology, Government of Gujarat. His research interests include networked sliding mode control, control of multi-agent systems and its applications. Dr. Mehta received the Pedagogical Innovation Award from Gujarat Technological University (GTU) in 2014, and Dewang Mehta National Education Award in 2018. He is a senior member of the IEEE, member of the IEEE Industrial Electronics Society (IES) and Control System Society (CSS), and a life member of the Indian Society for Technical Education (ISTE), Institute of Engineers India (IEI), and Systems Society of India (SSI).

Acronyms

Abbreviations

2DOFSFJ	2-Degree-Of-Freedom Serial Flexible Joint
DFSMC	Discrete-time First-order Sliding Mode Control
DHLMAS	Discrete Heterogeneous Linear Multi-Agent System
DHMAS	Discrete Homogeneous Multi-Agent System
DHSM	Discrete-time Higher order Sliding Mode
DMAS	Discrete Multi-Agent System
DOF	Degree Of Freedom
DSMC	Discrete-time Sliding Mode Control
DSTA	Discrete Super-Twisting Algorithm
Eqn	Equation
FA-1	Follower Agent-1
FA-2	Follower Agent-2
FA-3	Follower Agent-3
FA-4	Follower Agent-4
FA-5	Follower Agent-5
LQ	Linear Quadratic
LTI	Linear Time Invariant
MAS	Multi-Agent System
RD-2	Relative Degree-2
VSS	Variable Structure System

Symbols

$x(t)$	State vector of individual agent system in continuous-time domain
$u(t)$	Control input of individual agent system in continuous-time domain
$y(t)$	Output in continuous-time domain

$x(k)$	State vector of individual agent system in discrete-time domain
$u(k)$	Control input of individual agent system in discrete-time domain
$y(k)$	Output in discrete-time domain
Ψ,σ,M_s	User-defined gain constants
$\zeta_1,\ \zeta_2$	User-defined gain constants
G	System matrix in continuous-time domain
H	Input matrix in continuous-time domain
E	System matrix in discrete-time domain
F	Input matrix in discrete-time domain
C	Output matrix
$\tilde{E}$	Individual agent's system matrix in discrete-time domain
$\tilde{F}$	Individual agent's input matrix in discrete-time domain
$\bar{E}$	Global multi-agent system matrix in discrete-time domain
$\bar{F}$	Global multi-agent input matrix in discrete-time domain
$x_0(k)$	State vector of the leader agent in discrete-time domain
$S(t)$	Sliding variable in continuous-time domain
k_s	User-defined gain constant
σ_s	Sliding gain
$\lambda, \mathcal{F}, \rho$	User-defined constant
$S(k)$	Sliding variable in discrete-time domain
δ_0, ω_0	Positive offset
τ	Sampling time
r	Relative degree
$D(k)$	Slow time-varying disturbance signal in discrete-time domain
$h_1,\ h_2$	User-defined gain
$u_{eq}(k)$	Equivalent control in discrete-time domain
$\bar{G}$	Weighted diagraph
V	Vertex nodes
$\bar{\mathcal{E}}$	Set of edges
$\mathcal{A}$	Adjacency matrix
a_{ij}	Adjacency gain for communication between *i*th and *j*th agents
$\mathcal{N}$	Set of agents
$\tilde{D}$	In- or out-degree matrix
$\tilde{L}$	Laplacian matrix
a_{i0}	Adjacency gain for leader connected follower agents
$\tilde{\mathcal{B}}$	Diagonal matrix gives information of connecting agents with the leader
$\Gamma(.)$	Piecewise constant switching of graph
p	Switching instants for graph
$\tilde{\mathcal{D}}_i(k)$	Disturbance acting on *i*th agent in discrete-time domain
$\bar{\mathcal{D}}(k)$	Disturbance acting on global DMAS in discrete-time domain
$\otimes$	Kronecker product
$\bar{\delta}_i(k)$	Local neighborhood consensus error in discrete-time domain

$\bar{e}_i(k_p^i)$	Event-triggered-based local neighborhood consensus error in discrete-time domain
$\bar{\mu}_i(k)$	Local neighborhood consensus error for heterogeneous agent in discrete-time domain
$\bar{\Lambda}(k)$	Global consensus error in discrete-time domain
$\bar{\gamma}(k_p)$	Event-triggered-based global consensus error in discrete-time domain
$\bar{\varphi}(k)$	Global consensus error for heterogeneous leader-following DMAS in discrete-time domain
τ_i	Sampling time for the *i*th agent
λ_i, ϑ_i	User-defined controller gain for *i*th agent
$\bar{\lambda}, \bar{\vartheta}$	User-defined controller gain for global DMAS
Φ_i	Sliding mode band for the *i*th agent using Power rate reaching law
Δ_i	Quasi-sliding mode band for the *i*th agent using Gao's reaching law
$\mathcal{T}$	Transpose of the matrix
$\sigma_{s_i}^{\mathcal{T}}$	Sliding gain for *i*th agent
$k^\star$	Positive integer
$\bar{\mathcal{V}}(k)$	Lyapunov function in discrete-time domain
$\tilde{S}_i(k)$	Sliding variable for the *i*th agent in discrete-time domain
$\tilde{S}(k)$	Global sliding variable for DMAS in discrete-time domain
sgn	Signum function
$\bar{\alpha}$	Parameter constant for Lyapunov stability analysis
β_i	Upper bound of disturbance
η_i	User-defined constant for Power rate reaching law
θ	Pitch position of 2-DOF helicopter system
$\dot{\theta}$	Pitch velocity of 2-DOF helicopter system
$\ddot{\theta}$	Pitch acceleration of 2-DOF helicopter system
ψ	Yaw position angle of 2-DOF helicopter system
$\dot{\psi}$	Yaw velocity of 2-DOF helicopter system
$\ddot{\psi}$	Yaw acceleration of 2-DOF helicopter system
$\bigcup$	Union
$u_0(k)$	Control input of the leader agent in discrete-time domain
I	Idempotent matrix
N	Number of follower agents

List of Figures

List of Tables

Chapter 1
Introduction

1.1 Background of Multi-agent System

A Multi-Agent System (MAS) as shown in Fig. 1.1 is a combination of multiple interacting intelligent agents. The multi-agent system can also be defined as a group of agents which may be homogeneous, heterogeneous, or autonomous in nature and can communicate with each other to solve complex problems that are beyond the scope of an individual agent. The word "agent" illustrates a simple system dynamics and it may be a drone, wheeled mobile robot, helicopter system, etc. The agent may have capabilities such as autonomous behavior, communication with other agents, target-oriented behavior, and adaptation to changes in the environment, making it possible to bring new features to applications [1]. There are various control methodologies available in the literature for the coordination of the agents in a multi-agent system. Among them, the cooperative control method is a widely used technique for coordination among the agents.

1.2 Cooperative Control of Multi-agent System

The cooperative control of MAS has been widely researched area since last few years due to its broad applications such as **autonomous underwater vehicles** [2–4], **unmanned aerial vehicles** especially for military usage [5–7], **mobile robots** [8–12], **satellite clusters** [13, 14], **renewable energy** [15, 16], and many more unexplored area. Traditionally, a common centralized digital computer was employed to collect the information from the other system/agent for controlling the multi-agent system. However, it increases the complexity due to the communication network and also prone to failure due to system delay or corrupted data which deteriorates the performance of the MAS. To overcome these problems, distributed cooperative

K. Patel and A. Mehta, *Discrete-Time Sliding Mode Protocols for Discrete Multi-Agent System*, Studies in Systems, Decision and Control 303,
https://doi.org/10.1007/978-981-15-6311-9_1

Fig. 1.1 Multi-agent system representation

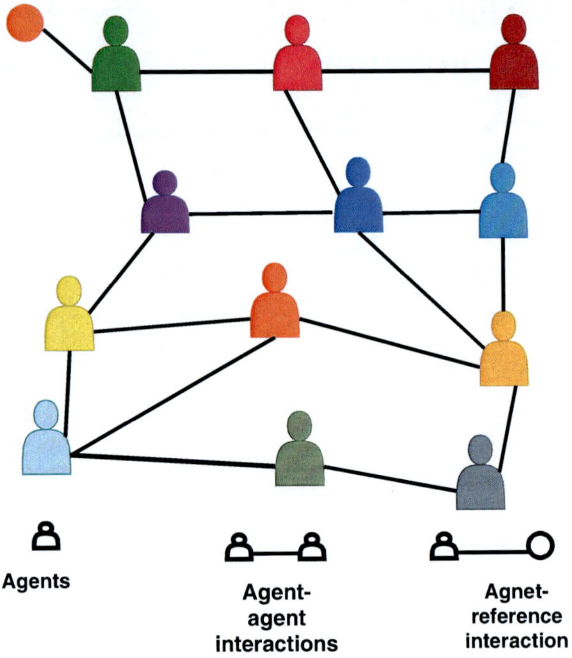

control is proposed in literature. The theme of distributed controllers lies in the nature. The collective behaviors of animal groups in nature have shown that distributed decisions made by each individual for its own position, direction, and speed of motion can make the whole group behave like a single entity, which has its own rules of motion and decision-making power. The examples of such a collective behavior can be seen in bird flocking [17], fire flies [18], schools of fish [19], and neuron firing as shown in Fig. 1.2. Inspired by these natural collective behaviors of animals, scientists and engineers got encouraged to develop distributed cooperative control for a group of systems by exchanging their information with neighbors (Fig. 1.2).

Earlier, the controllers were designed for coordination in a centralized structure in which a common digital computer is employed to collect the information from the other agent. It increases the complexity if one of the systems is delayed or corrupted. It also requires large bandwidth and processing time. To tackle this type of problems, distributed coordination control is proposed in which multiple agents in the same network can pass their information to neighbor agent and try to adjust each other for reaching a common global goal. Due to the availability of digital controller, sensor, and actuator, it is easier for the each agent to collect data from the network, to do their own task and also to interact with neighbor agents. Many practical and theoretical challenges are involved in cooperative control of multi-agent systems such as instead of a single system we have a system of subsystems which need to communicate together while the communication bandwidths are limited and it is also a difficult task to determine which agents to communicate at each time

(a) Fireflies [20] (b) Neuron firing [21]

(c) Flocking of birds [22] (d) Fish schooling [23]

Fig. 1.2 Cooperative behavior of multi-agent system in nature

and what to communicate. Moreover, there is a compromise between individual's goals and the team goal. In multi-agent system, there are a number of research problems that have resulted in the development of many useful tools and theories. To achieve this, many cooperative control algorithms for MAS are developed such as **formation** [24–26], **flocking** [27–29], **rendezvous** [30, 31], **consensus** [32, 33], etc. Among them, the consensus algorithm is widely researched and developed algorithm in recent years. The consensus approach of cooperative control algorithm provides a systematic framework for the design of multi-agent distributed controllers with general linear agent dynamics and dynamics with uncertainties.

Cooperative control of multi-agent systems covers a wide range of applications such as **autonomous underwater vehicles** [2–4], **unmanned aerial vehicles** especially for military usage[5–7], **mobile robots** [8–12], **satellite clusters** [13, 14], **renewable energy** [15, 16], and many more unexplored areas.

1.3 Consensus Algorithm for Cooperative Control

What is consensus? In the English dictionary, the consensus means to arrive at a general agreement in the interest of all concerns or agents. Consensus is the fundamental problem of cooperative control for MAS. The consensus of MAS can be divided into two types: The first one is the leaderless consensus [34] in which individual agent updates its information with the neighboring agent and agrees upon a certain degree of average value among them. The second one is the leader-following consensus [35] in which all the follower agents continually monitor the leader information and try to adjust with each other. Among these two, the leader-following consensus is vastly operated.

1.4 Literature Review of Consensus Algorithms

The literature survey on consensus as shown in Fig. 1.3 is spread over various domains and methodologies. A brief review of the same is discussed as follows.

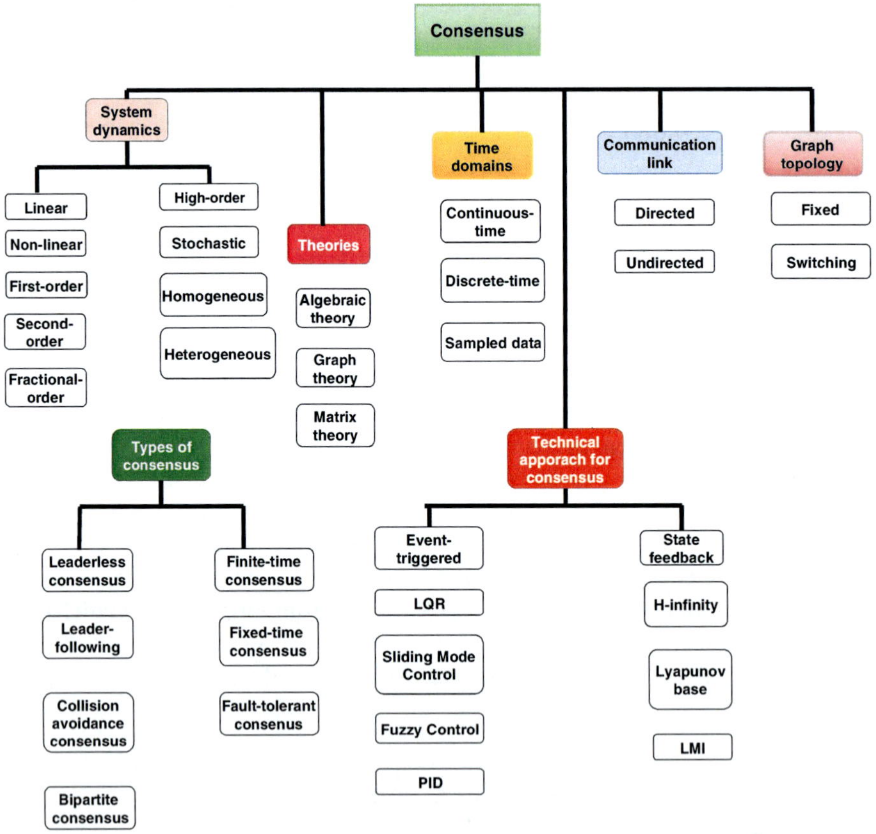

Fig. 1.3 Literature survey on consensus algorithms

1.4.1 System Dynamics

Generally, the agent dynamics can be classified into two categories, linear and nonlinear systems. Earlier, the linear system was more popular due to its easiness from both theoretical study and implementation point of view [36]. However, in the real-world problems, one has to pay more focus to inherent nonlinear characteristics of the dynamics [37]. The first-order dynamics containing the position information of the agent which is very simple and considered as primary system dynamics of MAS [38]. Hence, many researchers have focused on algorithm development for the first-order consensus of MAS [39–41]. Simultaneously, researchers and academicians studied consensus using second-order, higher order, and fractional-order dynamics of the multi-agent system. The authors in the paper [42, 43] discussed the second-order consensus for MAS. They developed the consensus protocol using graph theory and derived the stability condition for the consensus of MAS. C. Hua et al. [44] designed output higher order consensus for the leader-follower network of nonlinear multi-agent system under a fixed directed graph. Y. Ji et al. [45] proposed consensus of fractional-order MAS and derived fractional Lyapunov direct method for the asymptotic consensus. In a most practical scenario, the information received by the agents from their respective neighbor agents is contaminated by multiple noises and time delays, and hence it is a challenging task in such conditions to achieve consensus, so Y. Zhang et al. [46] designed the protocol for the stochastic consensus of MAS. B. Kaviarasan et al. [47] also designed the non-fragile consensus protocol for stochastic MAS with input time delay constraint.

In reference to the above literature review, MAS is mainly considered as a homogeneous system which means that agents should have the same dynamics with identical parameters and model configuration. But in certain conditions, it is difficult to implement homogeneous MAS to cooperatively accomplish the task, e.g., in the military mission of rescuing requires coordination between ground vehicle and the unmanned aerial vehicle. The research of heterogeneous MAS explores a further range in the field of military and civilian applications. Dun et al. [48] proposed the leader-following consensus for heterogeneous MAS with communication constraint.

1.4.2 Graph Theory and Communication Link

In literature, the communication among the agents in MAS is represented by algebraic theory, graph theory, and matrix theory [49–51]. Among all the theories, the graph theory has been used widely [52–54]. The communication topologies can be generally divided into two types, namely, (i) fixed graph topology and (ii) switching graph topology. Various authors [55–59] have proposed the consensus algorithms for fixed graph topology. Also, many authors [60–62] have proposed consensus protocols for the switching graph topology in which the communication topology changes dynamically.

1.4.3 Time Domains

In literature, the consensus algorithms have been developed in different time domains such as continuous-time, discrete-time, and sampled-data format. It is observed that most of the system dynamics are present in continuous-time domain for real-time operation such as the motion of the human body, autonomous quad-copter, and so on and hence the researchers have proposed consensus protocols in continuous-time [63]. On the other side, due to the recent development of digital technology and processors, it becomes interesting to develop the consensus algorithms for MAS in the discrete-time domain [64]. The sampled-data control consists of continuous-time system dynamics and discrete-time controllers. The authors in the paper [65] have proposed sample-data-based protocol for the multi-agent system with packet loss and other communication irregularities.

As shown in Fig. 1.3, the consensus algorithms for MAS have been proposed in both continuous-time domain and discrete-time domain as follows.

1.4.4 Consensus Algorithms for Continuous MAS

Different researchers and academicians around the globe [66–72] have given their insightful thought for different consensus methodologies of MAS. Y. Liu et al. [73] designed average consensus output feedback control using H_∞ approach. J. Wei et al. [74] proposed an average consensus using state feedback protocol with communication delays and also derived the sufficient condition for the convergence. Z. Liu et al. [75] investigated the average consensus problem of a multi-agent system with directed fixed communication topology using an event-triggered protocol. J. Fu et al. [76] designed a method for the leaderless exponential consensus of MAS using LMI technique. It is more crucial to opt for the leader-following consensus compared to leaderless consensus as the leader provides the path for reference. C.Q. ma et al. [77] studied the leader-following consensus of MAS using the Lyapunov approach with fixed and switching graph topology. J. Ma et al. [78] designed the leader-following first-order and second-order consensus using the linear quadratic regulator (LQR) method for MAS. B. Mu et al. [79] proposed a distributed LQR protocol for consensus of heterogeneous MAS and validated experimentally. C. Hua et al. [80] designed an adaptive consensus protocol for the leader-following network of nonlinear MAS. A. Shariati et al. [81] designed a leader-following consensus protocol using PID technique and derived sufficient condition for the consensus using LMI method. They applied this consensus protocol for the remote operation application of MAS. G. Wen et al. [82] proposed an observer-based consensus protocol for the leader-following consensus of fractional-order MAS. In order to save the communication bandwidth and power resources, many authors [83–85] opted for an event-triggered based leader-following consensus in which control updated only if triggering condition is satisfied. Few authors [86, 87] also investigated the self-triggered control which computes the next sampling or event ahead of time for the consensus of MAS. Up till now, the proposed protocols achieved the consensus asymptotically not in finite time and also not robust against model uncertainties and active disturbances present in the MAS. To deal with these problems, C. Ren et al. [88] designed the conventional sliding mode control for the leader-following consensus for second-order nonlinear MAS using a directed graph. Y. Jiang et al. [89] designed the robust integral sliding mode for the leader-following consensus of higher order MAS with fixed and switching topologies under the influence of disturbances. Nair et al. [90] designed an event-triggered-based integral sliding mode for the leader-following consensus of first-order MAS with uncertainties. In order to avoid conflict among agents, many scientists have studied for the collision avoidance approach for the consensus of MAS. J. Yan et al. [91] solved the collision avoidance problem for the consensus of MAS using a receding horizon approach and designed an optimal control algorithm. S. Dubay et al. [92] proposed MPC-based collision avoidance of multiple quad-copters trying to reach consensus. Apart from this, fault in DMAS can occur due to many reasons such as unavoidable disturbance in the system, sensor failure, actuator failure, and communication failure. Such problems can affect the performance of DMAS. Hence, many authors [93–95] have solved this problem using the fault-tolerant method. And also P. Yang et al. [96] proposed the consensus protocol for the leader-following network along with the disturbance, actuator faults and designed fault-tolerant method to compensate the effect of fault in the DMAS.

The recent development of digital communication technology enables researchers to design distributed cooperative control method in the discrete-time domain. In the next section, the consensus algorithms for DMAS in discrete-time domain is presented.

1.4.5 Consensus Algorithms for Discrete MAS

F. Wang. [97] proposed an average consensus protocol for Discrete Multi-Agent System (DMAS) . G. Gu et al. [98] designed the average consensus protocol using state feedback technique and discussed about the gain margin effect in the consensus of DMAS. W.Hou et al. [99] proposed a leaderless consensus for DMAS with delay constraint. Z.Ma et al. [100] and X. Xu et al. [101] developed observer-based consensus protocol for the leader-following DMAS and discrete-time Lyapunov method is used for the stability of the proposed protocol. Y. Wang et al. [102] developed a consensus protocol in which time-varying gain is applied to minimize the noise effect for the leader-following linear DMAS. J. Wu et al. [103] proposed the leader-following consensus protocol for nonlinear DMAS with minimal usage of communication equipment for fixed interaction graph topology. F. Wang et al. [104] proposed the leader-following consensus protocol using matrix theory with switching topology for second-order DMAS. M. Wyrwas et al. [105] designed a novel consensus protocol for the leader-following fractional-order DMAS and also derived the sufficient condition for the consensus. Y. Zhang et al. [106] discussed the higher order DMAS consensus protocol with switching topology. H. Zhao et al. [107] designed a distributed protocol for the leader tracking follower for the DMAS using LMI method. D. wang et al. [108] introduced an event-triggered consensus protocol for the leader-following heterogeneous DMAS with bounded delay. In this paper, first-order and second-order dynamics which are heterogeneous in nature are considered for the consensus. Mahmoud et al. [109] investigated H_∞ consensus protocol for the leader-following DMAS. L. Gao et al. [110] designed a control protocol using the Riccati equation for the leader-following consensus of DMAS and also derived sufficient condition using graph theory and matrix theory. M.S. Radenkovic et al. [111] discussed two novel adaptive cooperative distributed control algorithms for the leader-following DMAS with single direction graph topology. C. Ma et al. [112] developed the bipartite consensus protocol for the leader-following DMAS with the signed graph.

1.5 Organization of Book

The summary of the book is as follows:

- Chapter 1, presents brief introduction of multi-agent system and literature survey on different cooperative control algorithms for the consensus in continuous-time as well as discrete-time domain.
- In Chap. 2, the preliminaries of sliding mode control, particularly discrete-time sliding mode control and the graph theory used for developing the protocols are presented. The chapter also includes the preliminary results on discrete-time sliding mode protocols for the leader-following homogeneous Discrete Multi-Agent System (DMAS).
- In Chap. 3, two types of discrete-time sliding mode protocols for the leader-following consensus of homogeneous DMAS are proposed, namely, with the Gao's reaching law and the Power rate reaching law. The ultimate time for the consensus of homogeneous DMAS is also derived. Finally, the simulation and comparative results are presented to infer the consensus performance of both protocols for homogeneous DMAS comprising 2-DOF helicopters.
- In Chap. 4, the discrete-time sliding mode protocols presented in the previous chapter are extended for consensus of DMAS with the switching graph topology.
- Chapter 5 presents higher order DSM protocols for the leader-following consensus of homogeneous DMAS. The consensus performance of proposed protocols is evaluated experimentally on the laboratory setup comprising 2-DOF flexible joint robotic arms.
- Chapter 6 presents an event-triggered higher order DSM protocol using DSTA for the leader-following consensus of homogeneous 2-DOF robotic arms.

- Chapter 7 presents the higher order DSM protocol for heterogeneous DMAS having different dynamics. The protocol is experimentally validated on a setup comprising the 2-DOF robotic system with rigid joint and flexible joint.
- Chapter 8 concludes the monograph contribution with future challenges and the scope of this topic.

References

1. Fang, L.: Cooperative control of multi-agent systems with information flow constraints. Ph.D. thesis, University of Notre Dame (2006)
2. Yoon, S., Qiao, C.: Cooperative search and survey using autonomous underwater vehicles (auvs). IEEE Trans. Parallel Distrib. Syst. **22**, 364–379 (2011)
3. Simetti, E., Casalino, G.: Manipulation and transportation with cooperative underwater vehicle manipulator systems. IEEE J. Ocean. Eng. **42**, 782–799 (2017)
4. Chen, M., Zhu, D.: A novel cooperative hunting algorithm for inhomogeneous multiple autonomous underwater vehicles. IEEE Access **6**, 7818–7828 (2018)
5. Kim, S., Seo, S.: Cooperative unmanned autonomous vehicle control for spatially secure group communications. IEEE J. Select. Areas Commun. **30**, 870–882 (2012)
6. Li, P., Yu, X., Zhang, Y.: Fault-tolerant cooperative control for multiple uavs based on ude and model following smcâAŐAŐ this work was supported in part by natural sciences and engineering research council of canada, in part by the national natural science foundation of china under grants 61403407, 61573282, and 61603130. IFAC-PapersOnLine **51**(24) 447–452 (2018). 10th IFAC Symposium on Fault Detection, Supervision and Safety for Technical Processes SAFEPROCESS 2018
7. Huang, S., Teo, R.S.H., Liu, W.: Distributed cooperative avoidance control for multi-unmanned aerial vehicles. Actuators **8**, 1 (2018)
8. Savkin, A.V.: Coordinated collective motion of groups of autonomous mobile robots: analysis of vicsek's model. IEEE Trans. Autom. Control **49**, 981–982 (2004)
9. Rezaee, H., Abdollahi, F.: A decentralized cooperative control scheme with obstacle avoidance for a team of mobile robots. IEEE Trans. Indust. Electron. **61**, 347–354 (2014)
10. Du, H., Wen, G., Cheng, Y., He, Y., Jia, R.: Distributed finite-time cooperative control of multiple high-order nonholonomic mobile robots. IEEE Trans. Neural Netw. Learn. Syst. **28**, 2998–3006 (2017)
11. Mu, B., Chen, J., Shi, Y., Chang, Y.: Design and implementation of nonuniform sampling cooperative control on a group of two-wheeled mobile robots. IEEE Trans. Indust. Electron. **64**, 5035–5044 (2017)
12. Kamel, M.A., Yu, X., Zhang, Y.: Fault-tolerant cooperative control design of multiple wheeled mobile robots. IEEE Trans. Control Syst. Technol. **26**, 756–764 (2018)
13. Zhang, H., Gurfil, P.: Cooperative orbital control of multiple satellites via consensus. IEEE Trans. Aerosp. Electron. Syst. **54**, 2171–2188 (2018)
14. Zhang, H., Gurfil, P.: Cooperative control of multiple satellites via consensus. In: 2016 24th Mediterranean Conference on Control and Automation (MED), IEEE (2016)
15. Park, S.-H., Choi, J.-Y., Won, D.-J.: Cooperative control between the distributed energy resources in ac/dc hybrid microgrid. ISGT **2014**, 1–5 (2014)
16. Deblecker, O., Stevanoni, C., Vallée, F.: Cooperative control of multi-functional inverters for renewable energy integration and power quality compensation in micro-grids. In: 2016 International Symposium on Power Electronics, Electrical Drives, Automation and Motion (SPEEDAM), pp. 1051–1058 (2016)
17. Reynolds, C.W.: Flocks, herds and schools: a distributed behavioral model. ACM SIGGRAPH Comput. Graph. **21**, 25–34 (1987)

18. Bojic, I.: Firefly-inspired synchronization in multi-agent systems (extended abstract) (2012)
19. Ghosh, S., Lee, J.: Equivalent conditions for uniform asymptotic consensus among distributed agents. In: Proceedings of the 2010 American Control Conference, pp. 4821–4826 (2010)
20. Shoot, F.: Incredible photos of fireflies and tips on how to make your own (2017)
21. Bosgraaf, L.: Molecular shots, inc, of groningen, the netherlands, for "firing neurons" (2010)
22. Trowbridge, C.C.: On the origin of the flocking habit of migratory birds (1914)
23. Selayar Resort, Schooling fish Indonesia (2017)
24. Fax, J.A., Murray, R.M.: Information flow and cooperative control of vehicle formations. IEEE Trans. Autom. Control **49**, 1465–1476 (2004)
25. Ren, W.: Consensus strategies for cooperative control of vehicle formations. IET Control Theory Appl. **1**, 505–512 (2007)
26. Tomic, I., Milonidis, E., Halikias, G.D.: Lqr distributed cooperative control of a formation of low-speed experimental uavs. In: 2016 UKACC 11th International Conference on Control (CONTROL), pp. 1–6 (2016)
27. Chuang, Y., Huang, Y.R., D'Orsogna, M.R., Bertozzi, A.L.: Multi-vehicle flocking: Scalability of cooperative control algorithms using pairwise potentials. In: Proceedings 2007 IEEE International Conference on Robotics and Automation, pp. 2292–2299 (2007)
28. La, H.M., Sheng, W.: Flocking control algorithms for multiple agents in cluttered and noisy environments. In: Bio-Inspired Self-Organizing Robotic Systems, pp. 53–79. Springer, Berlin (2011)
29. Zhao, W., Chu, H., Zhang, M., Sun, T., Guo, L.: Flocking control of fixed-wing uavs with cooperative obstacle avoidance capability. IEEE Access **7**, 17798–17808 (2019)
30. Dimarogonas, D.V., Kyriakopoulos, K.J.: On the rendezvous problem for multiple nonholonomic agents. IEEE Trans. Autom. Control **52**, 916–922 (2007)
31. Lin, P., Ren, W., Wang, H., Al-Saggaf, U.M.: Multiagent rendezvous with shortest distance to convex regions with empty intersection: algorithms and experiments. IEEE Trans. Cybern. **49**, 1026–1034 (2019)
32. He, D., Shi, D., Sharma, R.: Consensus-based distributed cooperative control for microgrid voltage regulation and reactive power sharing. In: IEEE PES Innovative Smart Grid Technologies, Europe, pp. 1–6 (2014)
33. Islam, S., Liu, P.X., Saddik, A.E.: Consensus based distributed cooperative control for multiple miniature aerial vehicles with uncertainty. In: 2017 IEEE International Conference on Systems, Man, and Cybernetics (SMC), pp. 2477–2479 (2017)
34. Ren, W.: Distributed leaderless consensus algorithms for networkedeulerêĂŞlagrange systems. Int. J. Control **82**(11), 2137–2149 (2009)
35. Yan, H., Gao, L.: Leader-following consensus of multi-agent systems with and without time-delay. In: 2010 8th World Congress on Intelligent Control and Automation, pp. 4342–4347 (2010)
36. Cheng, L., Wang, Y., Ren, W., Hou, Z., Tan, M.: On convergence rate of leader-following consensus of linear multi-agent systems with communication noises. IEEE Trans. Autom. Control **61**, 3586–3592 (2016)
37. He, W., Zhang, B., Han, Q., Qian, F., Kurths, J., Cao, J.: Leader-following consensus of nonlinear multiagent systems with stochastic sampling. IEEE Trans. Cybern. **47**, 327–338 (2017)
38. Wang, Y., Xie, Y., Cheng, L.: Leader-following consensus of multi-agent systems with dynamic leader and measurement noises. In: 2017 36th Chinese Control Conference (CCC), pp. 8379–8384 (2017)
39. Zhenhua, W., Limei, L., Huanshui, Z.: First-order consensus problem for multi-agent systems with communication delay. In: 2015 34th Chinese Control Conference (CCC), pp. 1583–1587 (2015)
40. Zuo, Z., Yang, W., Tie, L., Meng, D.: Fixed-time consensus for multi-agent systems under directed and switching interaction topology. In: 2014 American Control Conference, pp. 5133–5138 (2014)

41. Sun, F., Tuo, M., Li, Y., Liu, F.: Finite-time consensus for first-order multi-agent systems with measurement noises. In: 2019 IEEE 3rd Information Technology, Networking, Electronic and Automation Control Conference (ITNEC), pp. 1818–1822 (2019)
42. Wang, C., Wang, X., Ji, H.: Leader-following consensus for a class of second-order nonlinear multi-agent systems. Syst. Control Lett. **89**, 61–65 (2016)
43. Zou, W., Xiang, Z., Ahn, C.K.: Mean square leader-following consensus of second-order nonlinear multiagent systems with noises and unmodeled dynamics. In: IEEE Transactions on Systems, Man, and Cybernetics: Systems, pp. 1–9 (2018)
44. Hua, C., Li, K., Guan, X.: Leader-following output consensus for high-order nonlinear multiagent systems. IEEE Trans. Autom. Control **64**, 1156–1161 (2019)
45. Ji, Y., Liu, Y., Guo, Y.: Consensus of fractional-order multi-agent systems via lmi approach. In: 2018 Chinese Control And Decision Conference (CCDC), pp. 907–912 (2018)
46. Zhang, Y., Li, R., Zhao, W., Huo, X.: Stochastic leader-following consensus of multi-agent systems with measurement noises and communication time-delays. Neurocomputing **282**, 136–145 (2018)
47. Kaviarasan, B., Sakthivel, R., Li, Y., Zhao, D., Ren, Y.: Non-fragile control protocol for finite-time consensus of stochastic multi-agent systems with input time-varying delay. Int. J. Mach. Learn, Cybern (2019)
48. Dun, Yu, M.: Leader-following consensus for heterogeneous multi-agent systems with bounded communication delays. In: 2016 14th International Conference on Control, Automation, Robotics and Vision (ICARCV), pp. 1–6 (2016)
49. Lee, H., Fujisaki, Y.: A characterization of algebraic connectivity of multi-agent systems with hierarchical network structure. IFAC-PapersOnLine **50**(1), 14206–14211 (2017). 20th IFAC World Congress
50. Feng, Z., Hu, G.: Distributed tracking control for multi-agent systems under two types of attacks. IFAC Proc. Vol. **47**(3), 5790–5795 (2014). 19th IFAC World Congress
51. Xia, H., Huang, T.-Z., Shao, J.-L., Yu, J.-Y.: Group consensus of multi-agent systems with communication delays. Neurocomputing **171**, 1666–1673 (2016)
52. Wang, F.-Y., Liu, Z.-X., Chen, Z.-Q.: A novel leader-following consensus of multi-agent systems with smart leader. Int. J. Control, Autom. Syst. **16**, 1483–1492 (2018)
53. Wang, Q., Sun, C.: Adaptive consensus of multiagent systems with unknown high-frequency gain signs under directed graphs. IEEE Transactions on Systems, Man, and Cybernetics: Systems, pp. 1–6 (2019)
54. Patel, K., Mehta, A.: Discrete-time sliding mode control for leader following discrete-time multi-agent system. In: IECON 2018 - 44th Annual Conference of the IEEE Industrial Electronics Society, pp. 2288–2292 (2018)
55. Xu, J., Li, T., Lum, K.Y., Xie, L.: Dynamic consensus and formation: fixed and switching topologies. IFAC Proc. Vol. **44**(1), 9188–9193 (2011). 18th IFAC World Congress
56. Wang, Z., Fan, Z., Liu, G.: Guaranteed performance consensus problems for nonlinear multi-agent systems with directed topologies. Int. J. Control, 1–11 (2018)
57. Pan, H., Yu, X., Guo, L.: Admissible leader-following consensus of fractional-order singular multi-agent system via observer-based protocol. IEEE Transactions on Circuits and Systems II: Express Briefs, p. 1 (2018)
58. Patel, K., Mehta, A.: Discrete higher order sliding mode protocol for leader-following consensus of heterogeneous discrete multi-agent system. In: Lecture Notes in Electrical Engineering, pp. 1–10. Springer, Singapore (2019)
59. Patel, K., Mehta, A.: Discrete-time event-triggered higher order sliding mode control for consensus of 2-dof robotic arms. Eur. J, Control (2020)
60. Munz, U., Papachristodoulou, A., Allgower, F.: Consensus in multi-agent systems with coupling delays and switching topology. IEEE Trans. Autom. Control **56**, 2976–2982 (2011)
61. Dai, J., Guo, G.: Event-triggered leader-following consensus for multi-agent systems with semi-markov switching topologies. Information Sciences **459**, 290–301 (2018)
62. Patel, K., Mehta, A.: Discrete-time sliding mode protocols for leader-following consensus of discrete multi-agent system with switching graph topology. Eur. J. Control **51**, 65–75 (2020)

63. Chen, Y., Shi, Y.: Consensus for linear multiagent systems with time-varying delays: A frequency domain perspective. IEEE Trans. Cybern. **47**, 2143–2150 (2017)
64. Yang, T., Meng, Z., Dimarogonas, D.V., Johansson, K.H.: Global consensus for discrete-time multi-agent systems with input saturation constraints. Automatica **50**(2), 499–506 (2014)
65. Xing, M., Deng, F., Hu, Z.: Sampled-data consensus for multiagent systems with time delays and packet losses. IEEE Transactions on Systems, Man, and Cybernetics: Systems, pp. 1–8 (2018)
66. Tao, L., Jifeng, Z.: Mean square average consensus of multi-agent systems with time-varying topologies and stochastic communication noises. In: 2008 27th Chinese Control Conference, pp. 552–556 (2008)
67. Zhu, W., Cheng, D.: Leader-following consensus of second-order agents with multiple time-varying delays. Automatica **46**(12), 1994–1999 (2010)
68. Mu, X., Xiao, X., Liu, K., Zhang, J.: Leader-following consensus of multi-agent systems with jointly connected topology using distributed adaptive protocols. J. Franklin Inst. **351**(12), 5399–5410 (2014)
69. Wang, Z.-X., Du, D.-J., Fei, M.-R.: Average consensus in directed networks of multi-agents with uncertain time-varying delays. Acta Autom. Sin. **40**(11), 2602–2608 (2014)
70. Parsegov, S., Polyakov, A., Shcherbakov, P.: Fixed-time consensus algorithm for multi-agent systems with integrator dynamics. IFAC Proc. Vol. **46**(27) 110–115 (2013). 4th IFAC Workshop on Distributed Estimation and Control in Networked Systems (2013)
71. Chen, S., Ho, D.W.C., Li, L., Liu, M.: Fault-tolerant consensus of multi-agent system with distributed adaptive protocol. IEEE Trans. Cybern. **45**, 2142–2155 (2015)
72. Wang, F., Yang, H.: Containment consensus of multi-agent systems with communication noises. In: Proceedings of 2016 Chinese Intelligent Systems Conference, pp. 189–197. Springer, Singapore (2016)
73. Liu, Y., Jia, Y., Du, J., Yuan, S.: Dynamic output feedback control for consensus of multi-agent systems: An hâĹd approach. In: 2009 American Control Conference, pp. 4470–4475 (2009)
74. Wei, J., Fang, H.: State feedback consensus for multi-agent system with multiple time-delays. JNW **8**, 1960–1966 (2013)
75. Liu, Z., Chen, Z., Yuan, Z.: Event-triggered average-consensus of multi-agent systems with weighted and direct topology. J. Syst. Sci. Compl. **25**, 845–855 (2012)
76. Fu, J., Wen, G., Yu, W., Huang, T., Cao, J.: Exponential consensus of multiagent systems with lipschitz nonlinearities using sampled-data information. IEEE Trans. Circuits Syst. I: Regul. Pap. **65**, 4363–4375 (2018)
77. Ma, C.-Q., Li, T., Zhang, J.-F.: Leader-following consensus control for multi-agent systems under measurement noises. IFAC Proc. Vol. **41**(2) 1528–1533 (2008). 17th IFAC World Congress
78. Ma, J., Zheng, Y., Wang, L.: Lqr-based optimal topology of leader-following consensus. Int. J. Robust Nonlinear Control **25**(17), 3404–3421
79. Mu, B., Shi, Y.: Distributed lqr consensus control for heterogeneous multiagent systems: Theory and experiments. IEEE/ASME Trans. Mechatron. **23**, 434–443 (2018)
80. Hua, C., You, X., Guan, X.: Adaptive leader-following consensus for second-order time-varying nonlinear multiagent systems. IEEE Trans. Cybern. **47**, 1532–1539 (2017)
81. Shariati, A., Tavakoli, M.: A descriptor approach to robust leader-following output consensus of uncertain multi-agent systems with delay. IEEE Trans. Autom. Control **62**, 5310–5317 (2017)
82. Wen, G., Zhang, Y., Peng, Z., Yu, Y., Rahmani, A.: Observer-based output consensus of leader-following fractional-order heterogeneous nonlinear multi-agent systems. Int. J. Control 1–9 (2019)
83. Li, H., Liao, X., Huang, T., Zhu, W.: Event-triggering sampling based leader-following consensus in second-order multi-agent systems. IEEE Trans. Autom. Control **60**, 1998–2003 (2015)
84. Xu, B., He, W., Ye, D.: Event-triggered consensus for general linear leader-following multi-agent systems under directed topologies. In: IECON 2018 - 44th Annual Conference of the IEEE Industrial Electronics Society, pp. 5971–5976 (2018)

85. Tan, X., Cao, J., Li, X.: Consensus of leader-following multiagent systems: a distributed event-triggered impulsive control strategy. IEEE Trans. Cybern. **49**, 792–801 (2019)
86. Wang, X., Su, H.: Self-triggered leader-following consensus of multi-agent systems with input time delay. Neurocomputing **330**, 70–77 (2019)
87. You, X., Hua, C., Guan, X.: Self-triggered leader-following consensus for high-order nonlinear multiagent systems via dynamic output feedback control. IEEE Trans. Cybern. **49**, 2002–2010 (2019)
88. Ren, C., Chen, C.L.P.: Sliding mode leader-following consensus controllers for second-order non-linear multi-agent systems. IET Control Theory Appl. **9**(10), 1544–1552 (2015)
89. Jiang, Y., Liu, J., Wang, S.: Robust integral sliding-mode consensus tracking for multi-agent systems with time-varying delay. Asian J. Control **18**, 224–235 (2016)
90. Nair, R.R., Behera, L., Kumar, S.: Event-triggered finite-time integral sliding mode controller for consensus-based formation of multirobot systems with disturbances. IEEE Trans. Control Syst. Technol. **27**, 39–47 (2019)
91. Yan, J., Guan, X.-P., Luo, X.-Y., Yang, X.: Consensus and trajectory planning with input constraints for multi-agent systems. Acta Autom. Sin. **38**(7), 1074–1082 (2012)
92. Dubay, S., Pan, Y.-J.: Distributed MPC based collision avoidance approach for consensus of multiple quadcopters. In: 2018 IEEE 14th International Conference on Control and Automation (ICCA), IEEE (2018)
93. Yang, P., Xu, M., Li, D., Liu, Z., Huang, Y.: Distributed fault tolerant consensus control for multi-agent system with actuator fault based on adaptive observer. Trans. Inst. Meas. Control **41**, 4207–4217 (2019)
94. Fattahi, M., Afshar, A.: Fault tolerant consensus control of multi-agent systems with switching topology. SN Appl. Sci. **1** (2019)
95. Chen, S., Ho, D.W.C., Li, L., Liu, M.: Fault-tolerant consensus of multi-agent system with distributed adaptive protocol. IEEE Trans. Cybern. **45**, 2142–2155 (2015)
96. Yang, P., Ma, B., Dong, Y., Liu, J.: Fault-tolerant consensus of leader-following multi-agent systems based on distributed fault estimation observer. Int. J. Control, Autom. Syst. **16**, 2354–2362 (2018)
97. Zhu, M., Martinez, S.: Discrete-time dynamic average consensus. Automatica **46**(2), 322–329 (2010)
98. Gu, G., Marinovici, L., Lewis, F.: Consensusability of discrete-time multi-agent systems under state feedback control. In: Proceedings of the 30th Chinese Control Conference, pp. 4889–4894 (2011)
99. Hou, W., Wu, Z., Fu, M., Zhang, H.: Constrained consensus of discrete-time multi-agent systems with time delay. Int. J. Syst. Sci. **49**(5), 947–953 (2018)
100. Ma, Z., Liu, Z., Chen, Z.: Modified leader-following consensus of time-delay multi-agent systems via sampled control and smart leader. Int. J. Control, Autom. Syst. **15**, 2526–2537 (2017)
101. Xu, X., Chen, S., Huang, W., Gao, L.: Leader-following consensus of discrete-time multi-agent systems with observer-based protocols. Neurocomputing **118**, 334–341 (2013)
102. Wang, Y., Cheng, L., Wang, H., Hou, Z., Tan, M., Yu, H.: Leader-following consensus of discrete-time linear multi-agent systems with communication noises. In: 2015 34th Chinese Control Conference (CCC), pp. 7010–7015 (2015)
103. Wu, J., Li, H., Chen, X.: Leader-following consensus of nonlinear discrete-time multi-agent systems with limited communication channel capacity. J. Franklin Inst. **354**(10), 4179–4195 (2017)
104. Wang, F., Liu, Z., Chen, Z.: Leader-following consensus of switched multi-agent systems with general second-order dynamics. In: 2017 Chinese Automation Congress (CAC), pp. 743–747 (2017)
105. Wyrwas, M., Mozyrska, D., Girejko, E.: Fractional discrete-time consensus models for single- and double-summator dynamics. Int. J. Syst. Sci. **49**(6), 1212–1225 (2018)
106. Zhang, Y., Li, S., Liao, L.: Consensus of high-order discrete-time multiagent systems with switching topology. IEEE Transactions on Systems, Man, and Cybernetics: Systems, pp. 1–10 (2018)

107. Zhao, H., Ren, W., Yuan, D., Chen, J.: Distributed discrete-time coordinated tracking with markovian switching topologies. Syst. Control Lett. **61**(7), 766–772 (2012)
108. Wang, D., Yu, M.: Leader-following consensus for heterogeneous multi-agent systems with bounded communication delays. In: 2016 14th International Conference on Control, Automation, Robotics and Vision (ICARCV), pp. 1–6 (2016)
109. Mahmoud, M.S., Khan, G.D.: Leader-following discrete consensus control of multi-agent systems with fixed and switching topologies. J. Franklin Inst. **352**(6), 2504–2525 (2015)
110. Gao, L., Cui, Y., Chen, W., Chen, W.: Leader-following consensus for discrete-time descriptor multi-agent systems with observer-based protocols. Trans. Inst. Meas. Control **38**(11), 1353–1364 (2016)
111. Radenkovic, M.S., Krstic, M.: Adaptive leader following in networks of discrete-time dynamical systems: algorithms and global convergence analysis. IEEE Trans. Control Netw. Syst. **6**, 324–337 (2019)
112. Ma, C., Zhao, W., Zhao, Y.: Bipartite consensus of discrete-time double-integrator multi-agent systems with measurement noise. J. Syst. Sci. Compl. **31**, 1525–1540 (2018)

Chapter 2
Preliminaries of Sliding Mode Control and Graph Theory

2.1 Review of Sliding Mode Control

A variable structure system with Sliding Mode Control (SMC) was first revealed to the world in 1964s in the Soviet Union by [1]. Since then, SMC has been developed into a general design method being examined for a wide range of systems including nonlinear systems and multi-input-multi-output systems. SMC is control method and a particular type of Variable Structure System (VSS) that alters the dynamics of a system by application of a discontinuous control signal. In the SMC, control law is no more continuous function of time. Instead, it can switch from one structure to other structure based on the present position of the system states. Thus, SMC is also called as a Variable Structure Control (VSC) method. The main properties of the VSC are the invariance to a bounded disturbance and insensitive to parameter variation.

In variable structure systems, the system is assumed to consist of continuous subsystems known as structures. These structures are changed depending on the states of the system. The states at which the structures change contribute to discontinuous surfaces in the phase planes. These surfaces are called as switching surfaces [2]. The following illustrative example is presented to understand the variable structure system with sliding mode control.

Consider two dynamical systems defined as follows:

$$\ddot{y} = -\zeta_1 y, \tag{2.1}$$

$$\ddot{y} = -\zeta_2 y,$$

where $0 < \zeta_2 < \zeta_1$.

The phase-plane trajectories of the systems defined in Eq. (2.1) are shown in Fig. 2.1. It can be revealed from the phase portraits that both systems are oscillatory and unstable in nature. However, when both systems are switched at a particular line

K. Patel and A. Mehta, *Discrete-Time Sliding Mode Protocols for Discrete Multi-Agent System*, Studies in Systems, Decision and Control 303,
https://doi.org/10.1007/978-981-15-6311-9_2

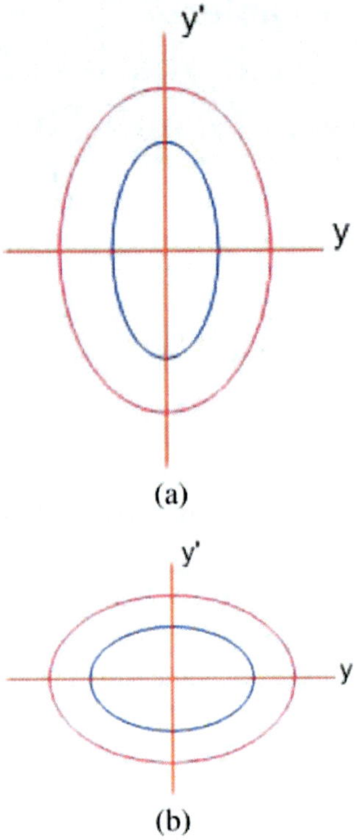

Fig. 2.1 Phase portrait of system (2.1) with **a** ζ_1 and **b** ζ_2, respectively

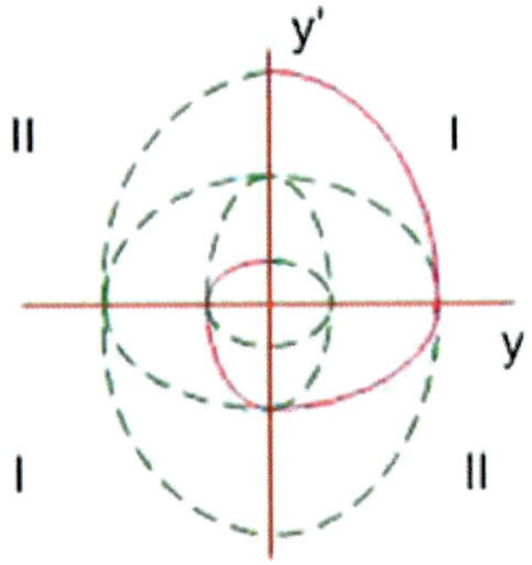

Fig. 2.2 Phase portrait of variable structure system

in the phase plane then the resulting system gives the stable response shown in the phase portrait Fig. 2.2.

In the above example, it can be seen from the phase portrait that the resulting VSS is asymptotic. However, it is also revealed by the researcher [3–5] that the VSS is switched at a particular line or plane then the closed-loop system provides not only finite-time convergence but also preserves the property of insensitivity to uncertainties and disturbances. The phenomenon can be well understood by the illustrative example given as follows. Consider the second-order linear time-invariant system as (Figs. 2.3, 2.4 and 2.5)

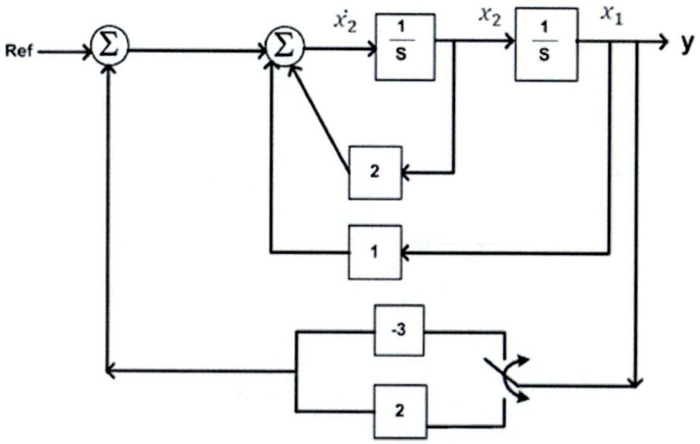

Fig. 2.3 Block diagram representation of VSC for second-order system

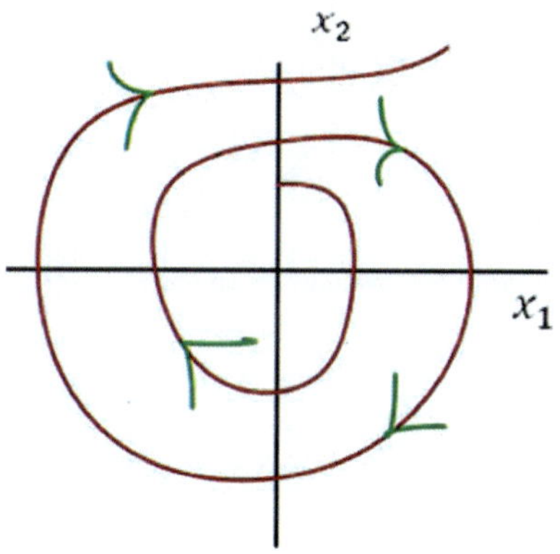

Fig. 2.4 Unstable phase portrait for (2.2) using gain = 3

$$\begin{bmatrix} \dot{x}_1 \\ \dot{x}_2 \end{bmatrix} = \begin{bmatrix} 0 & 1 \\ 1 & 2 \end{bmatrix} \begin{bmatrix} x_1 \\ x_2 \end{bmatrix} + \begin{bmatrix} 0 \\ 1 \end{bmatrix} u(t). \tag{2.2}$$

The overall behavior of the system is shown in Fig. 2.6. It is clearly observed from the figure that the system defined in Eq. (2.2) is an asymptotically stable variable structure system when the control is switched in accordance with a gain value which is −3 and 2, respectively. The typical form of switching surface of the aforementioned problem can be defined as

$$x_2 + \sigma x_1 = 0. \tag{2.3}$$

From Eq. (2.3), it is clear that system output response in sliding mode depends on the constant σ, and hence the system is insensitive to any external perturbation and parameter variations [6, 7]. This invariance property of VSS is extreme when controlling a time-varying plant or its parameter is continuously changing with respect to time [8].

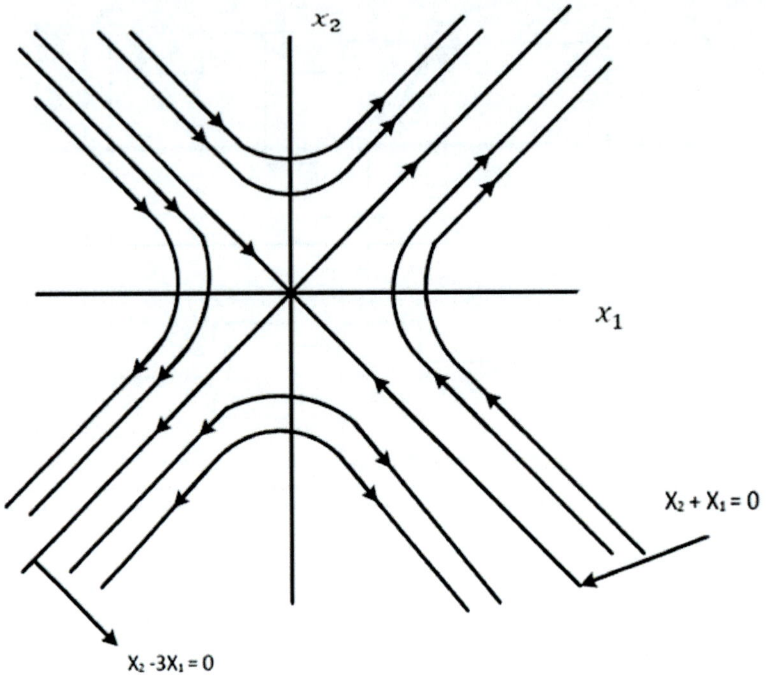

Fig. 2.5 Unstable phase portrait for (2.2) using gain = 2

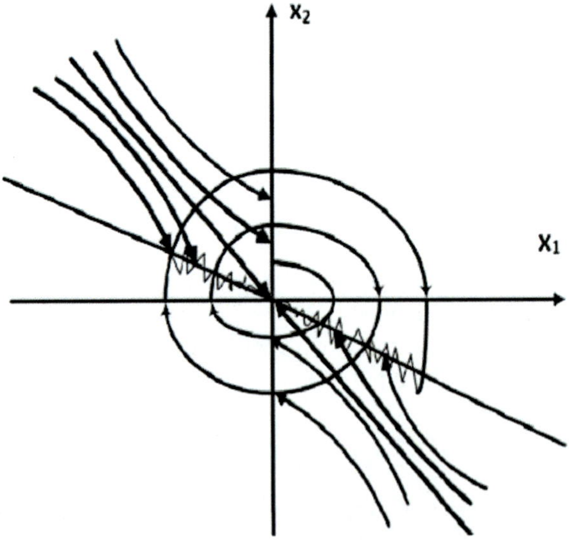

Fig. 2.6 Stable variable structure system consisting of two unstable structures

Further, it can be noted that different choices for the constant coefficient $'\sigma'$ in (2.3) result in different switching surfaces and hence get different system responses. This allows researchers to choose different control structures at different situations to meet the specified requirement in sliding mode.

The special nomenclature of VSC is given as a sliding mode control. Sliding mode control is a special kind of control method which alters the dynamics of a system by the application of discontinuous control signal which forces the system to "slide." From many control methods, SMC has emerged as one of the popular control techniques due to its robustness and simple structure. The theory and practical concepts of SMC have been applied to many challenging applications such as control of electrical motors [9, 10], robotics [11], unmanned vehicles [12], satellite navigation system [13], space application, and many more.

Due to the wide properties of SMC such as handling plant parameter variation and disturbances, it does not need exact modeling of the system. Further, it permits the decoupling of system motion into partial components that are lower in the dimensions. This reduces the complexity of the system and enhances system performance. The discontinuous control law of SMC can also be implemented using a conventional power converter. Due to such flexibility, SMC has been accepted by a wide range of scientific communities and also successfully applied to different applications. There are two schools of thought for designing and implementation of the sliding mode control, namely, (i) Continuous-time Sliding Mode Control (CSMC) and (ii) Discrete-time Sliding Mode Control (DSMC).

2.1.1 Continuous-Time Sliding Mode Control

The phase portrait of the system with sliding mode control law can be viewed as a combination of two phases, namely, 1. reaching phase and 2. sliding phase. The reaching phase drives the system state trajectories to the stable sliding manifold, while sliding phase drives the system states to equilibrium or origin. Figure 2.7 represents the sliding mode phase configuration with S as continuous-time sliding function given as follows:

$$S = \{x \in X | S(x, t) = 0\}. \tag{2.4}$$

In order to achieve the sliding mode, the following two properties must be ensured:

1. The system stability is strictly restricted to the sliding surface.
2. Sliding mode should occur within finite time.

The sufficient condition for the occurrence of sliding motion on sliding surface is given by

$$S\dot{S} < 0, \tag{2.5}$$

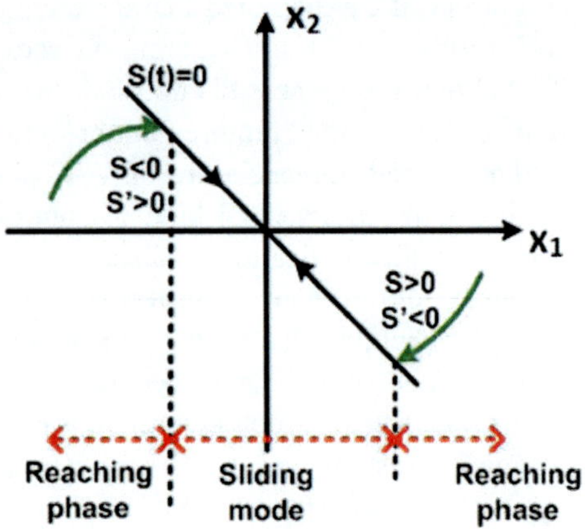

Fig. 2.7 Sliding mode phase

where S is the sliding surface and $\dot{S}$ is the rate of change of span from the sliding surface. The condition defined in (2.5) is called a reachability condition. The reachability condition is not sufficient for the occurrence of sliding motion on the sliding surface. The main lacuna of a condition defined in (2.5) gives asymptotical reach to the sliding surface and takes more amount of time to reach on the sliding surface. Hence, to overcome this problem another condition can be defined as follows:

$$S\dot{S} < -\Psi_s|S|, \quad \Psi_s < 0. \tag{2.6}$$

This condition is known as "Ψ_s- reachability" condition that ensures the finite-time convergence to $S = 0$.

As discussed previously, the design methodology for sliding mode control involves three subparts, namely, (i) sliding surface, (ii) reaching law, and (iii) switching control. To understand the entire process of sliding mode controller design, let us consider an LTI system as

$$\begin{aligned} \dot{X}(t) &= Ax(t) + Bu(t), \\ y(t) &= Cx(t). \end{aligned} \tag{2.7}$$

Now, sliding surface can be defined as

$$S(t) = \sigma_s x(t), \tag{2.8}$$

where σ_s is the sliding mode gain that can be calculated using different methods such as LQ, pole placement, Hurwitz, etc.

The briefing of proposed reaching laws in literature [14, 15] is

- Constant-proportional rate

$\dot{S} = -\lambda S - k_s sign(S), \lambda > 0$

- Power rate reaching law

$\dot{S} = -\lambda S - k_s|S|^{\mathcal{F}} sign(S), 0 < \mathcal{F} < 1.$

- Power rate exponential reaching law

$\dot{S} = -\frac{k_s}{N(S)} sign(S)$, where $N(s) = \delta_0 + (1 - \delta_0)e^{-\mathcal{F}|S|^{p_0}}$.

The dynamical equation of sliding function in Eq. (2.8) can be represented in the form of constant rate reaching law as follows:

$$\dot{S}(t) = -k_s sign(S), k_s > 0. \quad (2.9)$$

Hence, the control law is derived using (2.7) and (2.9) and can be written as

$$u(t) = -(\sigma_s B)^{-1}[A\sigma_s x(t) + k_s sign(S)]. \quad (2.10)$$

The main drawback of CSMC is that once the closed-loop system trajectories reach the sliding surface, the discontinuous control signal switches at very high frequency to achieve the sliding motion which is known as chattering in literature. The chattering in the closed loop is harmful as it may excite the unmodeled dynamics and also create excessive wear and tear on the actuator, particularly in a mechanical system. To alleviate the chattering, many algorithms have been proposed in the literature [16, 17]. In discrete time, SMC offers low switching frequencies because of the limitation of sampling frequency which makes it more preferable for implementation.

2.1.2 Discrete-Time Sliding Mode Control

Discrete-time sliding mode control was first introduced by C. Milosavljevic [18]. Subsequently, the other researchers [19–24] also extended their work in the discrete-time domain and designed DSMC for the different systems. In continuous-time sliding mode, once the closed-loop system is driven into the sliding mode, a discontinuous control signal switches with the infinite frequency while in the discrete-time sliding mode control is automatically constrained to the particular sampling frequency. It means that the control signal information is updated at the sampling instant only. In DSMC, the control input remains constant for the entire sampling period. So, in the case of DSMC, the states can never be on the sliding surface and move in a zig-zag form along the surface known as quasi-sliding mode. [25].

The DSMC is required to achieve the following performances:

1. Starting from any initial state, the trajectory will move monotonically toward the switching plane and crosses it in finite time.

2. Once the trajectory has crossed the switching plane, it will cross the plane again in every successive sampling period, resulting in a zig-zag motion about the switching plane.
3. The size of each successive zig-zag step is non-increasing and the trajectory stays within a specific band.

With the aforementioned discussion of DSMC, let us derive the DSMC algorithm using different reaching laws proposed in the literature. To design in a discrete domain, first, consider a continuous system as

$$\begin{aligned} \dot{X}(t) &= Ax(t) + Bu(t), \\ y(t) &= Cx(t), \end{aligned} \tag{2.11}$$

where $x \in \mathcal{R}^n$ is system state representation; $u \in \mathcal{R}^m$ is the control input; $y \in \mathcal{R}^p$ represents the system output; $A \in \mathcal{R}^{n\times n}$, $B \in \mathcal{R}^{n\times m}$, and $C \in \mathcal{R}^{p\times n}$ are the system, input, and output matrices, respectively.

Let the system defined in (2.11) discretized at the τ sampling interval given as

$$x(k+1) = Ex(k) + Fu(k), \tag{2.12}$$

$$y(k) = Gx(k). \tag{2.13}$$

where, $E = (exp)^{A\tau}$, $F = \int_0^{\tau}(exp)^{At}Bdt$.

As discussed earlier, the DSMC can be developed using the design of sliding surface and reaching law. The discrete-time sliding surface can be defined as

$$S(k) = \sigma_s x(k). \tag{2.14}$$

$$S(k+1) = \sigma_s x(k+1). \tag{2.15}$$

$$S(k+1) = \sigma_s(Ex(k) + Fu(k)). \tag{2.16}$$

Different researchers around the globe proposed reaching laws in the discrete-time domain as summarized below.

- Sarpturk's reaching law [26]:

$$\mid S(k+1) \mid < \mid S(k) \mid . \tag{2.17}$$

- Gao's reaching law [23]:

$$S(k+1) = (1-\lambda\tau)S(k) - \vartheta\tau sgn(S(k)), \tag{2.18}$$

where $\tau > 0$ is the sampling time, $\lambda > 0$, $\vartheta > 0$, and $(1-\lambda\tau) > 0$.

- Bartoszewicz's reaching law [24]:

$$S(k+1) = d(k) - d_0 + S_d(k+1), \tag{2.19}$$

where the unknown $d(k)$ is defined as $d_l \leq d(k) \leq d_u$ with d_u with d_l as lower bound and d_u as upper bound. $S_d(k)$ is a priori known function such that the following applies
- If $S(0) > 2\delta_d$ then

$$S_d(0) = S(0) \tag{2.20}$$

$$S_d(k)S_d(0) \geq, k \geq 0 \tag{2.21}$$

$$S_d(k) \geq 0, k \geq k^* \tag{2.22}$$

$$\mid S_d(k+1) \mid \leq \mid S_d(k) \mid -2\delta_d, k \leq k^*. \tag{2.23}$$

- Power rate reaching law [27]

$$S(k+1) = (1 - \lambda\tau)S(k) - \vartheta\tau|S(k)|^{\eta} sign(S(k)), \tag{2.24}$$

where $\lambda > 0$, $\vartheta > 0, 0 < \eta < 1$.

- Enhanced Power rate reaching law [28]

$$S(k+1) = (1 - \lambda\tau)\Omega(k)S(k) - \frac{q}{\Omega(k)}|S(k)|^{\frac{1}{2}} sign(S(k)), \tag{2.25}$$

where $\Omega(k) = \omega_0 + (1 - \omega_0)e^{-\rho|S(k)|p_0}, 0 < \omega_0 < 1, \rho > 0, \lambda > 0$.

If we consider Gao's reaching law, control $u(k)$ is obtained using (2.18), (2.16) and is given by

$$u(k) = -(\sigma_s F)^{-1}[E\sigma_s x(k) - (1 - \lambda\tau)S(k) + \vartheta\tau sgn(S(k))]. \tag{2.26}$$

Discrete-time sliding mode control has been shown to have many benefits including robustness with respect to internal and external matched uncertainties with known boundaries and system behavior specification by selecting the appropriate sliding manifold resulting in reduced order system dynamics in a sliding motion. However, it has continuously persisted a low-frequency chattering problem which might excite the unmodeled dynamics and deteriorate the system performance. This chattering problem has been taken care of by many researchers and they have introduced continuous approximation such as signum function, sat function [29, 30], etc. for a discontinuous term in the DFSMC. To solve this problem, higher order sliding mode control in the continuous-time domain is proposed by [31, 32]. Simultaneously, soham et al. [33] designed a discrete higher order SMC using switching-type reaching law with relative degree-1 and relative degree-2 concept. And it is revealed from the result that switching-type reaching law with relative degree-2 system gives better performance compared to relative degree-1. These results gave motivation to design a higher order SMC in discrete domain. Recently, in the paper [34] authors developed higher order DSMC for an uncertain LTI system in the presence of unmatched disturbances and successfully applied to the electromechanical system.

2.1.3 Discrete-Time Higher Order Sliding Mode Control

To understand the discrete-time higher order sliding mode control, it is required to understand the concept of relative degree of sliding mode control and particularly in the discrete-time domain.

The relative degree of discrete-time systems can be easily understood from the continuous-time concept [35].

Definition 2.1 ([33]) For a general discrete-time system defined as

$$\dot{x} = g_d(k, x(k), u(k)), \tag{2.27}$$

the output $y(k)$ has relative degree r if $y(k+r) = j_r(k, x(k), u(k))$ and $y(k+z) = j_z(k, x(k)) \forall 0 \leq z < r$, where u(k) is the control input and $y(k+P)$ illustrates the P unit delay of output y.

From above definition, first time the control input appears in the rth delay of the output $y(k)$ but not prior to that. Simply for LTI system (E, F, C), it can be defined as $CE^{z-1}F = 0, \ \ \forall z = 1 \ \ to \ \ (r-1)$ and $CE^rF \neq 0$.

Let us consider a discrete-time Linear Time-Invariant (LTI) system in the canonical form as

$$x_1(k+1) = E_{11}x_1(k) + E_{12}x_2(k)$$

$$x_2(k+1) = E_{21}x_1(k) + E_{22}x_2(k) + F_2u(k) + F_2D(k), \tag{2.28}$$

where $x_1(k) \in \mathbb{R}^{n-m}$, $x_2(k) \in \mathbb{R}^m$ are the states and $u(k) \in \mathbb{R}^m$ is the control input. The disturbance D with known bound and assumed as $||D(k)|| \leq \Lambda$. System matrix parameters $E_{11} \in \mathbb{R}^{(n-m)\times(n-m)}$, $E_{12} \in \mathbb{R}^{m\times(n-m)}$, $E_{21} \in \mathbb{R}^{m\times(n-m)}$, $E_{22} \in \mathbb{R}^{m\times m}$, and $F_2 \in \mathbb{R}^{m\times m}$ and assumed $\det(F_2) \neq 0$. Written in standard canonical form $x(k+1) = Ex(k) + F(u(k) + D(k))$ for LTI system, where $E = \begin{bmatrix} E_{11} & E_{12} \\ E_{21} & E_{22} \end{bmatrix}$ and $F = \begin{bmatrix} 0 \\ F_2 \end{bmatrix}$. For the system defined in (2.1.3), a relative degree-2 output can be written as

$$S_2(k) = \sigma_2^T x(k), \tag{2.29}$$

where $\sigma_2 \in \mathbb{R}^{m\times(n-m)}$.

$$S_2(k+1) = \sigma_2^T(Ex(k) + F(u(k) + D(k))). \tag{2.30}$$

In order to obtain the second-order sliding mode control u must appear in $k+2$ instant. Hence, σ_2^T is designed such that $\sigma_2^T F = 0$ and $\sigma_2^T EF \neq 0$. So (2.30) can be rewritten as

$$S_2(k+1) = \sigma_2^T(Ex(k)). \tag{2.31}$$

Now, control input $u(k)$ disappears from the system dynamics in Eq. (2.31), then we may write

$$S_2(k+2) = \sigma_2^T E^2 x(k) + \sigma_2^T EF(u(k) + D(k)). \tag{2.32}$$

Now control input $u(k)$ can be derived from Eq. (2.32).

- **Discrete-time super-twisting algorithm**

The aforementioned algorithm required a sliding function derivative for the implementation of the control law. However, Discrete-time Super-Twisting Algorithm (DSTA) has been designed for a system with relative degree-1 with respect to the sliding function and it does not require information of its derivative for implementation which is given as [36]

$$\begin{aligned} u_s(k) &= -h_1\sqrt{\mid S(k) \mid} sign(S(k)) + w(k) \\ w(k+1) &= w(k) - h_2 \tau sign(S(k)), \end{aligned} \tag{2.33}$$

where $h_1, h_2 > 0$, and τ is the sampling time.

$u(k)$ is combination of two terms,

$$u(k) = u_{eq}(k) + u_s(k), \tag{2.34}$$

where $u_{eq}(k)$ is equivalent control and $u_s(k)$ is switching control. For finding the equivalent control mentioned in above equation, let's consider $S(k+1) = 0$ in Eq. (2.16) and derive as

$$u_{eq}(k) = (\sigma_s F)^{-1}[\sigma_s Ex(k)]. \tag{2.35}$$

Remarks on DSMC

- Due to the digital arena, digital computers and microprocessors are available for implementation, so it is vital to design a system in a discrete domain.
- In continuous-time SMC, high-frequency chattering is produced which might cause a serious problem in the system. In contrast, DSMC offers low-frequency chattering which is more reliable and easy for practical implementation.
- However, DFSMC induces chattering and due to that system performance is degraded.

- Higher order DSMC attenuates chattering without compromising the original robust property of FDSMC. In a simple way, it hides the chattering inside the derivative of a sliding function.

As discussed in Sect. 1.1, in a multi-agent system, the agents communicate with the other neighbor agents for consensus. The communication topology in a multi-agent system is represented by the graph theory. The next section presents the preliminaries of the graph theory required for designing the sliding mode protocols for the consensus of MAS.

2.2 Preliminaries of Graph Theory

2.2.1 Graph Theory for Fixed Topology

Consider a weighted digraph $\bar{G} = (V, \bar{\mathcal{E}}, \mathcal{A})$ with a non-empty finite set of N nodes $V = \{v_1, v_2, ..., v_N\}$, a set of edges $\bar{\mathcal{E}} \subset V \times V$, and the associated adjacency matrix $\mathcal{A} = [a_{ij}] \in R^{N\times N}$ [37]. An edge rooted at node j and ended at node i is denoted by (v_j, v_i), which means that the information flows from node j to node i. The weight a_{ij} of edge (v_j, v_i) is positive, i.e., $a_{ij} > 0$ if $(v_j, v_i) \in \bar{\mathcal{E}}$, otherwise $a_{ij} = 0$. In this study, we assume that there are no repeated edges and no self-loops, i.e., $a_{ii} = 0, \quad \forall i \in \mathcal{N}$, where $\mathcal{N} = \{1, 2, ..., N\}$. If $(v_j, v_i) \in \bar{\mathcal{E}}$, then node j is called a neighbor of node i. The set of neighbors of node i is denoted by $\mathcal{N} = \{j|(v_j, v_i) \in \bar{\mathcal{E}}\}$. Define the in-degree matrix as $\tilde{D} = diag\{d_i\} \in \mathbb{R}^{N\times N}$ with $d_i = \sum_{j\in N_i} a_{ij}$ and the Laplacian matrix as $\tilde{L} = \tilde{D} - \mathcal{A}$. Hence, $\tilde{L}1_N = 0$. A graph is said to be directed if $(i, j) \in \bar{\mathcal{E}}$ if and only if the agent i can obtain information from the agent j. However, in the undirected graph $(i, j) \in \bar{\mathcal{E}}$ if and only if $(j, i) \in \bar{\mathcal{E}}$, but it is not necessary in the directed graph. Therefore, undirected graph can be regarded as the special digraph. A graph is said to be connected if every two vertices can be joined by a path. The digraph $\bar{G}$ has a globally node if and only if the Laplacian matrix $\tilde{L}$ of $\bar{G}$ has a simple zero eigenvalue with associated eigenvector $1_N = [1, 1, 1, ...1]^T$. If there is a node i_r such that there exists a directed path from the node i_r to every other nodes in a digraph, then it is said to have a spanning tree. A leader-following DMAS consisting of $N + 1$ agents labeled as $0, 1, 2, ..., N$. is considered. Without loss of generality, let agent 0 be the leader of DMAS and the other agents $1, 2, ..., N$ are the followers. The communications between followers are characterized by a digraph $\bar{G}$ with associated Laplacian matrix $\tilde{L}$. The digraph $\overline{G}$ characterizes the communications between the agents including all the followers and leader. The non-negative numbers a_{i0}, $i = 1, ..., N$ are used to describe the communications between followers and leader. If agent i gets the information from the leader, $a_{i0} > 0$, otherwise $a_{i0} = 0$. The diagonal matrix $\tilde{\mathcal{B}} = diag(a_{10}, .., a_{N0})$ is the communication weight matrix between followers and the leader [37–39].

2.2.2 Preliminaries of Graph Theory for Switching Topology

With switching topologies, the set $\bar{G} = \{\bar{G}_1, \bar{G}_2, ...\bar{G}_S\}$ where all subscripts of set $\bar{G}$ are presented as index set $\mathbb{S} = \{1, 2, ...S\}$ denotes the collection of all possible communication topologies. In this study, it is assumed that $\Gamma(k) : \mathbb{N} \to \mathbb{S}$ is a switching signal which describes the topologies switches at any time $t = k\tau$, except time interval $[k\tau, (k+1)\tau)$, where $\tau > 0$ is the sampling period.

Let's define a dynamic graph $\bar{G}_{\Gamma(k)} = (V, \bar{\mathcal{E}}_{\Gamma[k]}, \mathcal{A}_{\Gamma[k]}) \in \bar{G}$ at switching instant $\Gamma(k)$ which stand for the interaction topology of information exchange between N followers at time $k\tau$, where $V = 1, 2, ..., N$ and $\bar{\mathcal{E}}_{\Gamma[k]} \subseteq V \times V$. The weighted adjacency matrix of graph $\bar{G}_{\Gamma[k]}$ is denoted by $\mathcal{A}_{\Gamma[k]} = a_{ij}[k] \in R^{N\times N}$ with non-negative entries, where $a_{ij}[k] > 0$ if and only if $(j, i) \in \bar{\mathcal{E}}_{\Gamma[k]}$;otherwise, $a_{ij}[k] = 0$. Let $\tilde{L}_{\Gamma[k]} = [l_{ij}[k]] \in \mathbb{R}^{N\times N}$,$i, j = 0, 1..., N$ be the Laplacian matrix of the graph $\bar{G}_{\Gamma[k]}$. Let's define the degree matrix as $\tilde{D}_{\Gamma[k]} = diag\{d_i[k]\} \in \mathbb{R}^{N\times N}$ with $d[k] = \sum_{j\in N_i} a_{ij}[k]$ and the Laplacian matrix as $\tilde{L}_{\Gamma[k]} = \tilde{D}_{\Gamma[k]} - \mathcal{A}_{\Gamma[k]}$. Similarly the diagonal matrix $\tilde{\mathcal{B}}_{\Gamma[k]} = diag(a_{10}[k], .., a_{N0}[k])$ is the communication weight matrix between followers and the leader at time $k\tau$.

With this background of DSMC and graph theory, we proposed the preliminary result of DSM protocol for the leader-following consensus of DMAS in the next section. The discrete-time multi-agent system is configured with a fixed, undirected graph topology as a global system having one leader and other agents as followers.

2.3 Leader-Following Consensus of Homogeneous Discrete Multi-agent System

Consider the following identical linear discrete-time multi-agent system:

$$x_i(k+1) = \tilde{E}x_i(k) + \tilde{F}(u_i(k) + \tilde{\mathcal{D}}_i(k)) \ \forall_i \in N, \tag{2.36}$$

where $\tilde{E} \in \mathbb{R}^{n\times n}$ and $\tilde{F} \in \mathbb{R}^{n\times m}$ are the system matrix and input matrix, respectively. State vector $x_i(k) \in \mathbb{R}^n$ and the input vector $u_i(k) \in \mathbb{R}^m$, and $\tilde{\mathcal{D}}_i(k) \in \mathbb{R}^m$ is matched disturbance acting on ith agent system.

Assumption It is assumed that the matrix pair $(\tilde{E}, \tilde{F})$ for the ith system in Eq. (2.36) is controllable.

Assumption The communication topology of the DMAS under consideration is stationary and non-switching.

Assumption The disturbance $\tilde{\mathcal{D}}_i(k)$ is assumed to be bounded, where $\|\bar{\mathcal{D}}_i(k)\| \leq \beta_i$ with a known upper bound $\beta_i > 0$.

Lemma 2.1 ([41]) *If G satisfies Assumption 1, then* $(\tilde{L} + \tilde{\mathcal{B}})$ *is positive definite matrix.*

Definition 2.2 ([42]) Let $f(x) = R^n \to R^n$ represent a vector function and there exists $(\nabla_1, \nabla_2...., \nabla_n) \in R^n$ for $\forall \epsilon > 0$ such that $f(x)$ satisfies that $f_i(\epsilon^{\nabla_1} x_1, ..., \epsilon^{\nabla_n} x_n) = \epsilon^{k+\nabla_i} f_i(x)$ for $i = 1, 2, ...n$ where $k \geq max\{\nabla_i, i = 1, 2, ...n\}$, then $f(x)$ has homogeneous degree k associated with the dilation weight $(\nabla_1, \nabla_2...., \nabla_n)$ and system $\dot{x} = f(x)$ is called a homogeneous system.

The global homogeneous DMAS using Eq. (2.36) is derived as follows:

$$X(k+1) = (I_N \otimes \tilde{E})X(k) + (I_N \otimes \tilde{F})(u(k) + \tilde{\mathcal{D}}(k)), \tag{2.37}$$

$X(k) = [x_1(k), x_2(k)...x_N(k)]^{\mathcal{T}} \in \mathbb{R}^{nN}$ and the input vector $u(k) = [u_1(k), u_2(k)... u_N(k)]^{\mathcal{T}} \in \mathbb{R}^{mN}$, $\tilde{\mathcal{D}}(k) = [\bar{\mathcal{D}}_1(k), \bar{\mathcal{D}}_2(k), \bar{\mathcal{D}}_3(k), ...\bar{\mathcal{D}}_N(k)]^{\mathcal{T}} \in \mathbb{R}^{mN}$ matched disturbance vector acting on ith system.

We may rewrite Eq. (2.37) as

$$X(k+1) = \bar{E}X(k) + \bar{F}(u(k) + \bar{\mathcal{D}}(k)), \tag{2.38}$$

where $\bar{E} = (I_N \otimes \tilde{E})$, $\bar{F} = (I_N \otimes \tilde{F})$.

The dynamics of the leader is defined as

$$x_0(k+1) = \tilde{E}x_0(k), \tag{2.39}$$

where $x_0(k) \in \mathbb{R}^n$ is the state vector of the leader.

The goal is to design a robust DSM protocol for the consensus of homogeneous DMAS such that all the follower agents (2.38) achieved the consensus with the leader (2.39) in finite time using the neighborhood agent information.

It is assumed that the leader node can be observed from a small subset of nodes in graph $\bar{G}$. If ith agent is connected to the leader then this particular edge is said to exist with weighting gain $a_{i0} > 0$. The agent with $a_{i0} > 0$ is referred as pinning gain.

2.3.1 *Design of DSM Protocol for Homogeneous DMAS*

In this section, a distributed protocol for the leader-following consensus of homogeneous DMAS is designed using the discrete-time sliding mode control approach. Let us define the local neighborhood consensus error for the system defined in Eq. (2.36) as

$$\bar{\delta}_i(k) = \sum_{j \in N} a_{ij}[x_i(k) - x_j(k)] + a_{i0}[x_i(k) - x_0(k)]. \tag{2.40}$$

From Eq. (2.40), the consensus error in global form can be written as

$$\bar{\Lambda}(k) = ((I_N + \tilde{D} + \tilde{\mathcal{B}})^{-1})(\tilde{L} + \tilde{\mathcal{B}})) \otimes I_n)\tilde{x}, \tag{2.41}$$

where $\tilde{x} = X(k) - 1_N \otimes x_0(k) \in \mathbb{R}^{nN}$ and $((I_N + \tilde{D} + \tilde{\mathcal{B}})^{-1})$ is non-singular matrix. The eigenvalues of the weighted matrix $\tilde{\xi} = ((I_N + \tilde{D} + \tilde{\mathcal{B}})(\tilde{L} + \tilde{\mathcal{B}}))$ strictly remain inside the unit circle as per the Gershgorin circle criteria [43].

Definition 2.3 A complete circle $\mathbb{C}(\bar{c}_0, r_0)$ related to weighted matrix $\tilde{\xi}$ is a closed circle in the complex plane centered at $\bar{c}_0 \in \mathbb{R}$ and eigenvalues $\chi_i(\tilde{\xi}) \in \mathbb{C}(\bar{c}_0, r_0)$ for all $i = 1, 2, ..., N$.

So Eq. (2.41) is written as

$$\bar{\Lambda}(k) = (\tilde{\xi} \otimes I_n)\tilde{x}. \tag{2.42}$$

Substituting $\tilde{x}$, we may write

$$\bar{\Lambda}(k) = \bar{\Upsilon}(X(k) - 1_N \otimes x_0(k)), \tag{2.43}$$

where $(\tilde{\xi} \otimes I_n) = \bar{\Upsilon}$.

Let us define the sliding surface for the ith agent as

$$\tilde{S}_i(k) = \sigma_{s_i}^{\mathcal{T}} \bar{\delta}_i(k), \tag{2.44}$$

where $\sigma_{s_i}^{\mathcal{T}}$ is the sliding gain to be obtained using pole-placement approach [44].

Then for global system (2.38) sliding surface can be written as

$$\tilde{S}(k) = \sigma_s^{\mathcal{T}} \bar{\Lambda}(k), \tag{2.45}$$

where $\sigma_s^{\mathcal{T}} = I_N \otimes \sigma_{s_i}^{\mathcal{T}}$.

Using Eq. (2.43), we may write

$$\tilde{S}(k) = \sigma_s^{\mathcal{T}} \bar{\Upsilon}(X(k) - 1_N \otimes x_0(k)). \tag{2.46}$$

Further, advancing of Eq. (2.46)

$$\tilde{S}(k+1) = \sigma_s^{\mathcal{T}} \bar{\Upsilon}(X(k+1) - 1_N \otimes x_0(k+1)). \tag{2.47}$$

Substituting Eqs. (2.38) and (2.39) into Eq. (2.47), we may get

$$\tilde{S}(k+1) = \sigma_s^{\mathcal{T}} \bar{\Upsilon}(\bar{E}X(k) + \bar{F}(u(k) + \bar{\mathcal{D}}(k)) - 1_N \otimes \tilde{E}x_0(k)). \tag{2.48}$$

Now inspired from the reaching law defined in [23], let us define the consensus reaching law for the ith agent as

$$\tilde{S}_i(k+1) = (1 - \lambda_i \tau_i)\tilde{S}_i(k) - \vartheta_i \tau_i sgn(\tilde{S}_i(k)). \tag{2.49}$$

Then for global consensus reaching law for global homogeneous DMAS defined as

$$\tilde{S}(k+1) = (1 - \bar{\lambda}\tau)\tilde{S}(k) - \bar{\vartheta}\tau sgn(\tilde{S}(k)), \tag{2.50}$$

where $\bar{\vartheta} = [\vartheta_1, \vartheta_2, \vartheta_3, ...\vartheta_N] \in \mathbb{R}^{nN} > 0$, $\bar{\lambda} = [\lambda_1, \lambda_2, \lambda_3, ...\lambda_N] \in \mathbb{R}^{nN} > 0$, $0 < (1 - \bar{\lambda}\tau) < 1$.

Comparing Eqs. (2.48) and (2.50), the DSM protocol is derived as

$$\begin{aligned} u(k) = [-(\sigma_s^{\mathcal{T}}\bar{\Upsilon}\bar{F})^{-1}[\sigma_s^{\mathcal{T}}\bar{\Upsilon}\bar{E}X(k) + \sigma_s^{\mathcal{T}}\bar{\Upsilon}(-1_N \otimes \tilde{E}_0 x_0(k)) \\ -(1 - \bar{\lambda}\tau)\tilde{S}(k) + \bar{\vartheta}\tau sgn(\tilde{S}(k))]] - \bar{\mathcal{D}}(k). \end{aligned} \tag{2.51}$$

Using this protocol, the global consensus error considered as a global surface reaches to the sliding mode band in finite time. As discussed earlier, once the trajectory of consensus error reaches to the sliding surface, it moves in a zig-zag step and remained in a specified band called Quasi-Sliding Mode Band (QSMB) [23]. So the width of the QSMB for each agent (Δ_i) is defined using (2.49) for ith follower agent defined as

$$\Delta_i = \left(\frac{\vartheta_i \tau_i}{2 - \lambda_i \tau_i}\right). \tag{2.52}$$

Theorem 2.1 *The necessary condition for global stability of leader-following consensus using proposed protocol (2.51) is guaranteed if the eigenvalue of closed-loop dynamics of system matrix for homogeneous DMAS in (2.38) restricts to the bounded region* $\mathbb{C}(1, 1)$ *is*

$$[(I_{nN} - \bar{F}(\sigma_s^{\mathcal{T}}\bar{\Upsilon}\bar{F})\sigma_s^{\mathcal{T}})\bar{E}] < 1. \tag{2.53}$$

Proof To derive this closed-loop stability, consider system (2.38)

$$X(k+1) = \bar{E}X(k) + \bar{F}(u(k) + \bar{\mathcal{D}}(k)). \tag{2.54}$$

Substituting the control law $u(k)$ from Eq. (2.51) into Eq. (2.54), we may get

$$\begin{aligned} X(k+1) = \bar{E}X(k) - \bar{F}\big[(\sigma_s^{\mathcal{T}}\bar{\Upsilon}\tilde{F})]^{-1}[\sigma_s^{\mathcal{T}}\bar{\Upsilon}\bar{E}X(k) + (\sigma_s^{\mathcal{T}}\bar{\Upsilon} - 1_N \otimes \tilde{E}x_0(k)) \\ -(1 - \bar{\lambda}\tau)\tilde{S}(k) + \bar{\vartheta}\tau sgn(\tilde{S}(k))\big] + \bar{F}\bar{\mathcal{D}}(k). \end{aligned} \tag{2.55}$$

To find a sliding mode gain $\sigma_s^{\mathcal{T}}$ in the Eq. (2.55) is most difficult task. Thus, in the next, design procedure to obtain the sliding mode gain $\sigma_s^{\mathcal{T}}$ is given.

Design of sliding mode gain for DMAS

The sliding mode gain design for a sliding surface introduced here is mainly based on the pole-placement method for the global LTI DMAS (2.38) without any uncertainties is expressed by,

$$X(k+1) = \bar{E}X(k) + \bar{F}u(k). \tag{2.56}$$

The pair$(\bar{E}, \bar{F})$ is controllable, thus using pole-placement method a state feedback gain $\bar{K} = [K_1, K_2, ..K_N]_{mN \times nN}$ can be obtained by assigning nN eigenvalues $\left\{\chi_{11}, \chi_{21}, ..\chi_{n1}.., \chi_{12}, \chi_{22}, .., \chi_{n2}, ...\chi_{nN}\right\}$ for $(\bar{E} - \bar{F}\bar{K})$. To design $\sigma_s^{\mathcal{T}}$ in Eq. (2.38), the following conditions are considered.

Condition C$_1$: The matrix $(\bar{E} - \bar{F}\bar{K})$ has no eigenvalues same as matrix $\bar{E}$.
Condition C$_2$: The eigenvalues of matrix $(\bar{E} - \bar{F}\bar{K})$ are choosen to be stable remains inside the unit circle of complex plane and denoted as

$$\left\{\chi_{11}, \chi_{21}, ..\chi_{n1-m1}, \chi_{12}, \chi_{22}, .., \chi_{n2-m2}, ...\chi_{nN-mN}, \underbrace{\chi_{m1}, \chi_{m2}, \chi_{m3}, ...\chi_{mN}}_{m\ common\ \chi\ for\ N\ agents}\right\}, \tag{2.57}$$

with $\chi_{nN-mN} \neq \chi_{mN}$.
Condition C$_3$: The Matrix $(\bar{E} - \bar{F}\bar{K})$ is diagonalizable.
It has been proved that eigenvalues given as (2.57), where the number of the repeated eigenvalues χ is not greater than mN, thus

$$(\bar{E} - \bar{F}\bar{K})\bar{W} = \bar{W}\bar{J}, \tag{2.58}$$

where $\bar{W} = I_N \otimes \tilde{W}_{n \times n}$, $\bar{J} = I_N \otimes \tilde{J}_{n \times n}$, $\tilde{J} = diag[\chi_1, \chi_2, ...\chi_{n-m}, \underbrace{\chi ... \chi}_{m\chi s}]$ and $\bar{W}$ is non-singular. Further, to derived the sliding mode gain $\sigma_s^{\mathcal{T}}$, let's consider Eq. (2.58)

$$\bar{E}\bar{W} - \bar{F}\bar{K}\bar{W} = \chi\bar{W}, \tag{2.59}$$

Now multiplying both sides of Eq. (2.59) with $\sigma_s^{\mathcal{T}}$, we may get

$$\sigma_s^{\mathcal{T}}\bar{E}\bar{W} - \sigma_s^{\mathcal{T}}\bar{F}\bar{K}\bar{W} = \sigma_s^{\mathcal{T}}\chi\bar{W}, \tag{2.60}$$

Further,

$$\sigma_s^{\mathcal{T}}\bar{E}\bar{W} - \sigma_s^{\mathcal{T}}\chi\bar{W} = \sigma_s^{\mathcal{T}}\bar{F}\bar{K}\bar{W}, \tag{2.61}$$

Considereing $\sigma_s^{\mathcal{T}}\bar{F} = I_{mN}$, Eq. (2.61) is written as

$$\sigma_s^{\mathcal{T}} = \bar{K}(\bar{E} - \chi I_{nN})^{-1} \tag{2.62}$$

The global system matrix closed-loop dynamics written from Eq. (2.55) as

$$\bar{E}_{co} = [(I_{nN} - \bar{F}(\sigma_s^{\mathcal{T}}\Upsilon\bar{F})\sigma_s^{\mathcal{T}})\bar{E}] < 1. \tag{2.63}$$

Eigenvalue of Eq. (2.63) which lies on the bounded region $\mathbb{C}(1, 1)$ which is inside the unit circle.

It completes the proof.

2.3.2 Simulation Results

In this section, second-order system is considered as agent to show the efficacy of the proposed algorithm.

Consider multi-agent system with three follower nodes and one leader node. Leader node notation is given as 0 while follower node notations are given as $1, 2, and 3$, respectively. Their communication topology is shown in Fig. 2.8 and consider the system in (2.36) discretize at $\tau = 0.03$ sampling time interval, where $\tilde{E}$ is

$$\tilde{E} = \begin{bmatrix} 0.9324 & 0.1792 \\ 0 & 0.8607 \end{bmatrix}, \tilde{F} = \begin{bmatrix} -0.0122 \\ -0.0929 \end{bmatrix}. \tag{2.64}$$

The initial state of each agent is generated randomly in the span $[-1, 1]$ and the initial state of the leader node is $[0.1, 0.5]^T$, and the diagonal matrix $\tilde{D}$, adjacency matrix $\mathcal{A}$, pinning gain matrix $\tilde{\mathcal{B}}$, and Laplacian matrix $\tilde{L}$ are defined as

$$\tilde{D} = \begin{bmatrix} 2 & 0 & 0 \\ 0 & 2 & 0 \\ 0 & 0 & 2 \end{bmatrix}, \mathcal{A} = \begin{bmatrix} 0 & 1 & 1 \\ 1 & 0 & 1 \\ 1 & 1 & 0 \end{bmatrix}, \tilde{\mathcal{B}} = diag\{1, 1, 0\}, \tilde{L} = \begin{bmatrix} 2 & -1 & -1 \\ -1 & 2 & -1 \\ -1 & -1 & 2 \end{bmatrix}. \tag{2.65}$$

In order to check the robustness of the derived distributed DSM protocol for the global system, a slow varying disturbance is applied to the each DMAS with magnitude $\tilde{\mathcal{D}}_i(k) = 0.0002 * cos(0.86k)$. Gains for each agent ϑ_i and λ_i are chosen as 1 and 30, respectively. The sliding gain is calculated using the pole-placement

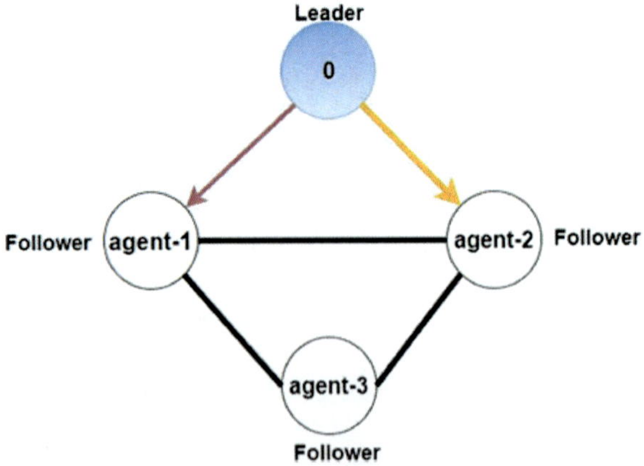

Fig. 2.8 Communication topology

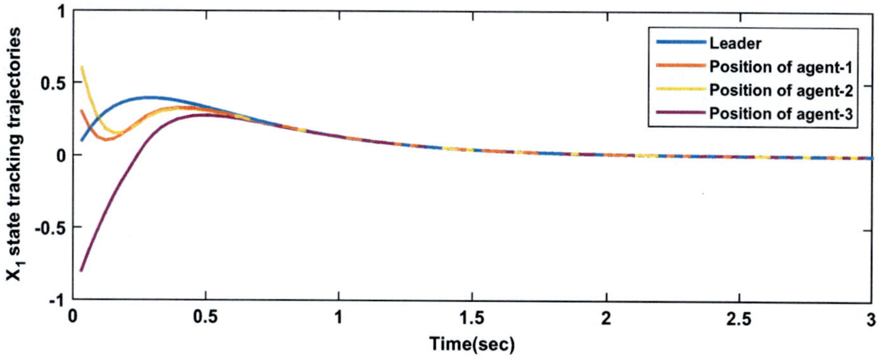

Fig. 2.9 Consensus of follower agent position with the leader position

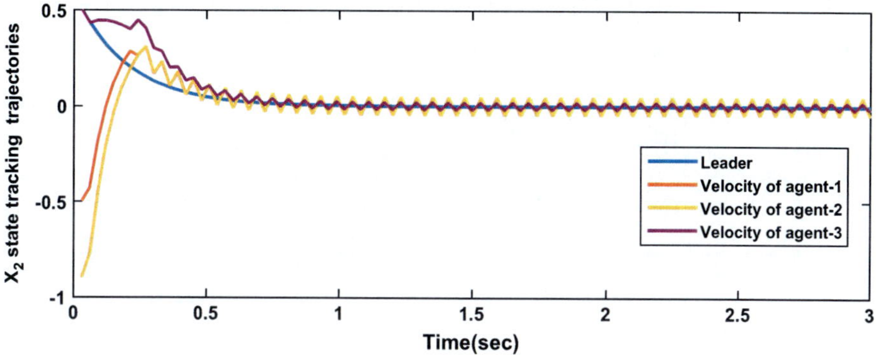

Fig. 2.10 Consensus of agent velocity with the leader velocity

method which comes out to be

$$\sigma_{s_i}^{\mathcal{T}} = \left[-1.5609 \; -2.4698\right]. \tag{2.66}$$

Figure 2.9 shows that position of each agent of DMAS follows the position of leader and the consensus begins in finite time.

Figure 2.10 shows that velocity of each agent of DMAS follows the velocity of leader and the consensus begins in finite time.

Figure 2.11 shows the control law (u) for the individual agent of DMAS.

Figure 2.12 shows the sliding variable of each agent of DMAS. The state trajectory crosses first time sliding surface from any initial condition in finite time and remains in QSMB and value of this varied from -0.04 to +0.04 calculated using Eq. (2.52).

Figures 2.13 and 2.14 present the consensus error of position and velocity states which eventually goes to zero in finite time.

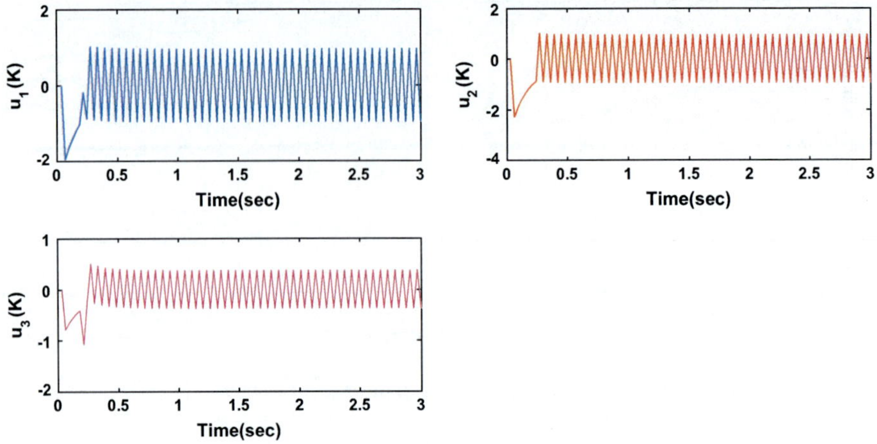

Fig. 2.11 Control effort of individual agent

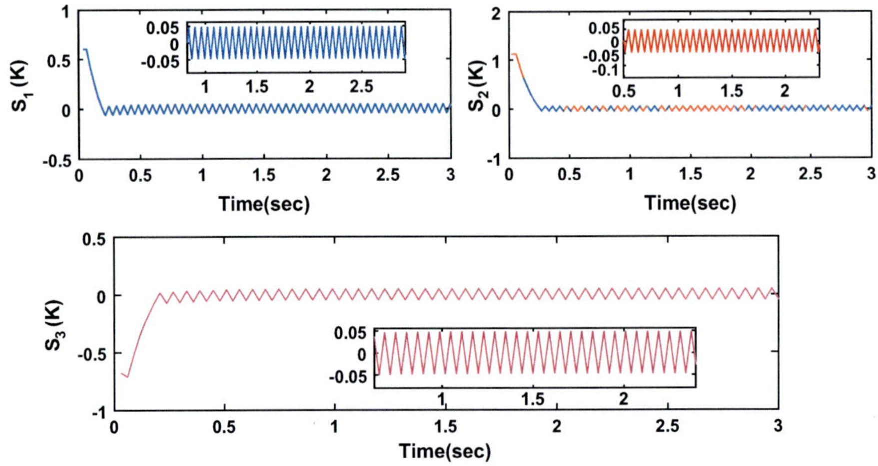

Fig. 2.12 Sliding variable of each agent

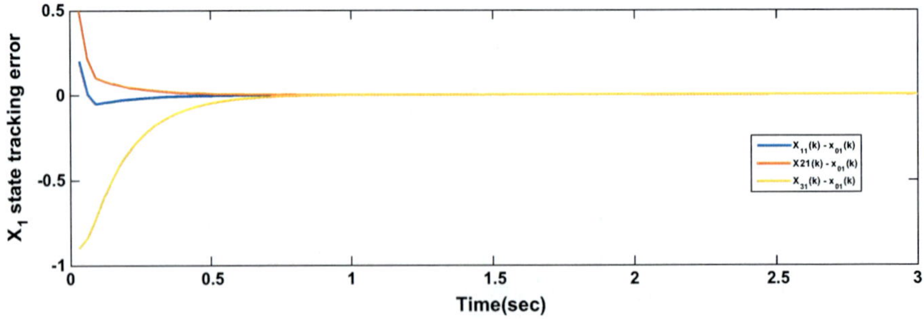

Fig. 2.13 Error between the position of agent and the leader

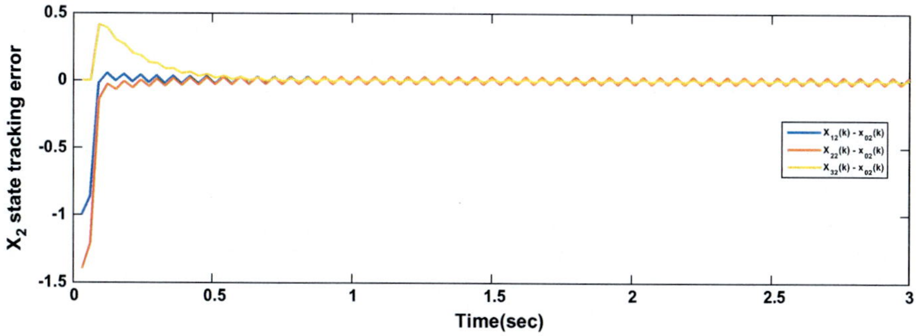

Fig. 2.14 Error between the velocity of follower agent and the leader

2.3.3 Conclusion

In this chapter, a new algorithm for leader-following consensus with discrete-time sliding mode control has been investigated. The proposed DSM protocol achieved the leader-following consensus in finite time. The necessary condition for global stability using the proposed DSM protocol has been derived. Finally, one simulation example is presented to show the efficacy of the proposed protocol.

References

1. Emel'yanov, SV., Utkin, V..: Stability of motion of a class of variable structure control systems. Izv. AN SSSR, Tech. Cyber. **2**, 140–142 (1964)
2. Utkin, V.: Variable structure systems with sliding modes. IEEE Trans. Autom. Control **22**, 212–222 (1977)
3. Utkin, V.I.: Sliding mode control design principles and applications to electric drives. IEEE Trans. Indust. Electron. **40**, 23–36 (1993)
4. Young, K., Özgüner, U. (eds.): Variable Structure Systems, Sliding Mode and Nonlinear Control. Springer, London (1999)
5. Furuta, K., Pan, Y.: Variable structure control with sliding sector. Automatica **36**, 211–228 (2000)
6. Emel'yanov, S.V., Utkin, V.I.: Synthesis of variable structure control systems with a discontinuous switching function, AN SSSR. Tech. Cyber **1**, 182–186 (1964)
7. Bartolini, G., Zolezzi, T.: Asymptotic linearization of uncertain systems by variable structure control. Syst. Control Lett. **10**(2), 111–117 (1988)
8. Thorp, J.S., Barmish, B.R.: On guaranteed stability of uncertain linear systems via linear control. J. Optim. Theory Appl. **35**, 559–579 (1981)
9. Izosimov, D., Utkin, V.: Sliding mode control of electric motors. IFAC Proc. Vol., **14**(2), 2059 – 2066 (1981). 8th IFAC World Congress on Control Science and Technology for the Progress of Society, Kyoto, Japan, 24-28 August 1981
10. Betin, F., Capolino, G.: Sliding mode control for electrical drives. In: 2015 IEEE International Electric Machines Drives Conference (IEMDC), pp. 1043–1048 (2015)
11. Zhang, B., Yang, X., Zhao, D., Spurgeon, S.K., Yan, X.: Sliding mode control for nonlinear manipulator systems. IFAC-PapersOnLine **50**(1), 5127–5132 (2017). 20th IFAC World Congress

12. Lee, S., Lee, J., Lee, S., Choi, H., Kim, Y., Kim, S., Suk, J.: Sliding mode guidance and control for uav carrier landing. IEEE Trans. Aerosp. Electron. Syst. **55**, 951–966 (2019)
13. Yu, C., Xie, X.: Dynamic sliding mode-based attitude stabilisation control of satellites with angular velocity and control constraints. Trans. Inst. Meas. Control **41**(4), 934–941 (2019)
14. Mehta, A., Bandyopadhyay, B.: Frequency-Shaped and Observer-Based Discrete-time Sliding Mode Control. Springer, India (2015)
15. Devika, K.B., Thomas, S.: Power rate exponential reaching law for enhanced performance of sliding mode control. Int. J. Control, Autom. Syst. **15**, 2636–2645 (2017)
16. Park, M.-H., Kim, K.-S.: Chattering reduction in the position control of induction motor using the sliding mode. IEEE Trans. Power Electron. **6**, 317–325 (1991)
17. Kawamura, A., Itoh, H., Sakamoto, K.: Chattering reduction of disturbance observer based sliding mode control. IEEE Trans. Ind. Appl. **30**, 456–461 (1994)
18. Milosavljevic, C.: General condition for the existance of a quasi-sliding mode in the switching hyperplane in discrete varaible structure system. Autom. Remote Control **46**, 307–314 (1985)
19. Corless, M., Manela, J.: Control of uncertain discrete-time systems. In: 1986 American Control Conference, pp. 515–520 (1986)
20. Furuta, K.: Sliding mode control of a discrete system. Syst. Control Lett. **14**, 145–152 (1990)
21. Chan, C.: Robust discrete-time sliding mode controller. Syst. Control Lett. **23**(5), 371–374 (1994)
22. Bartolini, G., Ferrara, A., Utkin, V.: Adaptive sliding mode control in discrete-time systems. Automatica **31**(5), 769–773 (1995)
23. Gao, W., Wang, Y., Homaifa, A.: Discrete-time variable structure control systems. IEEE Trans. Ind. Electron. **42**, 117–122 (1995)
24. Bartoszewicz, A.: Discrete-time quasi-sliding-mode control strategies. IEEE Trans. Ind. Electron. **45**, 633–637 (1998)
25. Bartoszewicz, A., Lesniewski, P.: Reaching law approach to the sliding mode control of periodic review inventory systems. IEEE Trans. Autom. Sci. Eng. **11**, 810–817 (2014)
26. Sarpturk, S., Istefanopulos, Y., Kaynak, O.: On the stability of discrete-time sliding mode control systems. IEEE Trans. Autom. Control **32**, 930–932 (1987)
27. Devika, K.B., Thomas, S.: Power rate exponential reaching law for enhanced performance of sliding mode control. Int. J. Control, Autom. Syst. **15**, 2636–2645 (2017)
28. Ma, H., Li, Y., Xiong, Z.: Discrete-time sliding-mode control with enhanced power reaching law. IEEE Trans. Ind. Electron. **66**, 4629–4638 (2019)
29. Edwards, C.: Sliding Mode Control. CRC Press (1998)
30. Suleiman, H.U., Muâazu, M.B., Zarma, T.A., Salawudeen, A.T., Thomas, S., Galadima, A.A.: Methods of chattering reduction in sliding mode control: a case study of ball and plate system. In: 2018 IEEE 7th International Conference on Adaptive Science Technology (ICAST), pp. 1–8 (2018)
31. Levant, A.: Sliding order and sliding accuracy in sliding mode control. Int. J. Control **58**, 1247–1263 (1993)
32. Khan, M.K., Spurgeon, S.K.: Second order sliding mode control of coupled tanks. IFAC Proc. Vol. **38**(1), 872–877 (2005)
33. Chakrabarty, S., Bandyopadhyay, B., Bartoszewicz, A.: Discrete-time sliding mode control with outputs of relative degree more than one. In: Recent Developments in Sliding Mode Control Theory and Applications, InTech (2017)
34. Sharma, N.K., Janardhanan, S.: Discrete-time higher-order sliding mode control of systems with unmatched uncertainty. Int. J. Robust Nonlinear Control **29**, 135–152 (2018)
35. Levant, A.: Principles of 2-sliding mode design. Automatica **43**(4), 576–586 (2007)
36. Salgado, I., Kamal, S., Bandyopadhyay, B., Chairez, I., Fridman, L.: Control of discrete time systems based on recurrent super-twisting-like algorithm. ISA Trans. **64**, 47–55 (2016)
37. Patel, K., Mehta, A.: Discrete-time sliding mode control for leader following discrete-time multi-agent system. In: IECON 2018 - 44th Annual Conference of the IEEE Industrial Electronics Society, pp. 2288–2292 (2018)

38. Patel, K., Mehta, A.: Discrete higher order sliding mode protocol for leader-following consensus of heterogeneous discrete multi-agent system. In: Lecture Notes in Electrical Engineering, pp. 1–10. Springer, Singapore (2019)
39. Patel, K., Mehta, A.: Discrete-time event-triggered higher order sliding mode control for consensus of 2-dof robotic arms. Eur. J. Control (2020)
40. Patel, K., Mehta, A.: Discrete-time sliding mode protocols for leader-following consensus of discrete multi-agent system with switching graph topology. Eur. J. Control **51**, 65–75 (2020)
41. Ren, C., Chen, C.L.P.: Sliding mode leader-following consensus controllers for second-order non-linear multi-agent systems. IET Control Theory Appl. **9**(10), 1544–1552 (2015)
42. Rosier, L.: Homogeneous lyapunov function for homogeneous continuous vector field. Syst. Control Lett. **19**, 467–473 (1992)
43. Varga, R.S.: Geršgorin and His Circles. Springer, Berlin (2004)
44. Nguyen, T., Miao, Z., Pan, Y., Amini, N., Ugrinovskii, V., James, M.R.: Pole placement approach to coherent passive reservoir engineering for storing quantum information. Control Theory Technol. **15**, 193–205 (2017)

Chapter 3
Discrete-Time Sliding Mode Protocols for Leader-Following Consensus of Homogeneous Discrete Multi-Agent System with Fixed Graph Topology

3.1 Introduction of the Discrete Leader-Following MAS

Consider the linear homogeneous DMAS comprises N agents having system dynamics of each agent as

$$x_i(k+1) = \tilde{E}x_i(k) + \tilde{F}(u_i(k) + \tilde{\mathcal{D}}_i(k)) \ \forall_i \in N, \tag{3.1}$$

where $i = 1, ..., N$, $\tilde{E} \in \mathbb{R}^{n\times n}$ is the system matrix, $\tilde{F} \in \mathbb{R}^{n\times m}$ is the input matrix of ith system, respectively. State vector $x_i(k) \in \mathbb{R}^n$ and the input vector $u_i(k) \in \mathbb{R}^m$, $\tilde{\mathcal{D}}_i(k) \in \mathbb{R}^m$ is matched disturbance acting on ith agent system.

Assumption $(\tilde{E}, \tilde{F})$ for ith agent system is controllable.

From Eq. (3.1), we may write the global homogeneous DMAS as

$$X(k+1) = (I_N \otimes \tilde{E})X(k) + (I_N \otimes \tilde{F})(u(k) + \tilde{\mathcal{D}}(k)), \tag{3.2}$$

where, $X(k) = [x_1(k), x_2(k)...x_N(k)]^{\mathcal{T}} \in \mathbb{R}^{nN}$, $u(k) = [u_1(k), u_2(k)...u_N(k)]^{\mathcal{T}} \in \mathbb{R}^{mN}$, $\tilde{\mathcal{D}}(k) = [\bar{\mathcal{D}}_1(k), \bar{\mathcal{D}}_2(k), \bar{\mathcal{D}}_3(k), ...\bar{\mathcal{D}}_N(k)]^{\mathcal{T}} \in \mathbb{R}^{mN}$ are system state vector, input, and disturbance of the global system, respectively. For generality, we may rewrite Eq. (3.2) as

$$X(k+1) = \bar{E}X(k) + \bar{F}(u(k) + \bar{\mathcal{D}}(k)), \tag{3.3}$$

where $\bar{E} = (I_N \otimes \tilde{E})$, $\bar{F} = (I_N \otimes \tilde{F})$.

Let us also consider the autonomous leader dynamics for leader-following consensus as

$$x_0(k+1) = \tilde{E}_0 x_0(k), \tag{3.4}$$

where $x_0(k) \in \mathbb{R}^n$ is the state vector of the leader.

K. Patel and A. Mehta, *Discrete-Time Sliding Mode Protocols for Discrete Multi-Agent System*, Studies in Systems, Decision and Control 303,
https://doi.org/10.1007/978-981-15-6311-9_3

Definition 3.1 The homogeneous DMAS system is said to achieve the consensus with leader in fixed time steps for $k \geq k^\star$

$$\|X(k) - x_0(k)\| = 0, \tag{3.5}$$

where $k^\star$ is positive integer.

Problem statement: To derive the number of steps required for global consensus of leader-following DMAS using the DSM protocols designed using Gao's reaching law and Power rate reaching law.

If there is an information linkage of ith agent to the leader then the weighted gain of the communication line of agents is $a_{i0} > 0$. The agent with $a_{i0} > 0$ is known as the pinning gain matrix[1–3].

The local neighborhood leader-follower network error is defined as

$$\bar{\delta}_i(k) = \sum_{j \in N} a_{ij}[x_i(k) - x_j(k)] + a_{i0}[x_i(k) - x_0(k)]. \tag{3.6}$$

From Eq. (3.6) and considering Lemma 3.1, the global consensus error can be written as

$$\bar{\Lambda}(k) = ((I_N + \tilde{D} + \tilde{\mathcal{B}})^{-1})(\tilde{L} + \tilde{\mathcal{B}})) \otimes I_n)\tilde{x}, \tag{3.7}$$

where $\tilde{x} = X(k) - 1_N \otimes x_0(k) \in \mathbb{R}^{nN}$. The eigenvalues of the weighted matrix $\tilde{\xi} = ((I_N + \tilde{D} + \tilde{\mathcal{B}})(\tilde{L} + \tilde{\mathcal{B}}))$ strictly remain inside the unit circle as per the Gershgorin circle criteria [4].

The global consensus error defined in Eq. (3.7) may be written in compressed form as

$$\bar{\Lambda}(k) = (\tilde{\xi} \otimes I_n)\tilde{x}. \tag{3.8}$$

Consider $(\tilde{\xi} \otimes I_n) = \bar{\Upsilon}$ and inserting the value of $\tilde{x}$, we can get

$$\bar{\Lambda}(k) = \bar{\Upsilon}(X(k) - 1_N \otimes x_0(k)), \tag{3.9}$$

$$\bar{\Lambda}(k) = \bar{\Upsilon}(X(k+1) - 1_N \otimes x_0(k+1)), \tag{3.10}$$

$$\bar{\Lambda}(k) = \bar{\Upsilon}(\bar{E}X(k) + \bar{F}(u(k) + \bar{\mathcal{D}}(k)) - 1_N \otimes \tilde{E}x_0(k)). \tag{3.11}$$

3.2 Analysis for Number of Steps Required for Global Consensus of DMAS Using the Protocol with the Gao's Reaching Law

In section, first, we estimate the fixed time steps required to achieve the consensus and then derived the sufficient condition for global stability of DMAS with the DSM protocol using Gao's reaching law.

Lemma 3.1 ([1]) *For the global DMAS (3.3) is said to achieve the global consensus with the leader agent in a fixed-time steps using protocol given as*

$$u(k) = [-(\sigma_s^{\mathcal{T}} \bar{\Upsilon} \bar{F})^{-1}[\sigma_s^{\mathcal{T}} \bar{\Upsilon} \bar{E} X(k) + \sigma_s^{\mathcal{T}} \bar{\Upsilon}(-1_N \otimes \tilde{E}_0 x_0(k)) - (1 - \bar{\lambda}\tau)\tilde{S}(k) + \bar{\vartheta}\tau sgn(\tilde{S}(k))]] \\ -\tilde{\mathcal{D}}(k), \tag{3.12}$$

where $(\sigma_s^{\mathcal{T}} \bar{\Upsilon} \bar{F})^{-1}$ *is non-singular,* $\sigma_s^{\mathcal{T}} = I_N \otimes \sigma_{s_i}^{\mathcal{T}}$.

Proof The sliding surface of ith agent is defined as

$$\tilde{S}_i(k) = \sigma_{s_i}^{\mathcal{T}} \bar{\delta}_i(k), \tag{3.13}$$

where $\sigma_{s_i}^{\mathcal{T}}$ is the sliding mode gains for ith agent to be obtained using the pole-placement approach ([5]) and τ is the sampling period of discrete-time system. From (3.13), the global sliding surface can be defined as

$$\tilde{S}(k) = \sigma_s^{\mathcal{T}} \bar{\Lambda}(k). \tag{3.14}$$

Using Eq. (3.9), we can get

$$\tilde{S}(k) = \sigma_s^{\mathcal{T}} \bar{\Upsilon}(X(k) - 1_N \otimes x_0(k)). \tag{3.15}$$

Further,

$$\tilde{S}(k+1) = \sigma_s^{\mathcal{T}} \bar{\Upsilon}(X(k+1) - 1_N \otimes x_0(k+1)). \tag{3.16}$$

Substituting Eqs. (3.3) and (3.4) into (3.16), we may get

$$\tilde{S}(k+1) = \sigma_s^{\mathcal{T}} \bar{\Upsilon}(\bar{E} X(k) + \bar{F}(u(k) + \tilde{\mathcal{D}}(k)) - 1_N \otimes \tilde{E}_0 x_0(k)). \tag{3.17}$$

Now using Gao's reaching law defined in ([6]), reaching law of ith agent is defined as

$$\tilde{S}_i(k+1) = (1 - \lambda_i \tau_i)\tilde{S}_i(k) - \vartheta_i \tau_i sgn(\tilde{S}_i(k)), \tag{3.18}$$

where τ_i is sampling period, $\lambda_i, \vartheta_i > 0$ are design parameters for the controller.

From Eq. (3.18), the global consensus reaching law for DMAS can be written as

$$\tilde{S}(k+1) = (1 - \bar{\lambda}\tau)\tilde{S}(k) - \bar{\vartheta}\tau sgn(\tilde{S}(k)), \tag{3.19}$$

where $\bar{\vartheta} = [\vartheta_1, \vartheta_2, \vartheta_3, ...\vartheta_N] \in \mathbb{R}^{nN} > 0$, $\bar{\lambda} = [\lambda_1, \lambda_2, \lambda_3, ...\lambda_N] \in \mathbb{R}^{nN} > 0$, $0 < (1 - \bar{\lambda}\tau) < 1$.

Comparing Eqs. (3.17) and (3.19), we may get

$$\sigma_s^{\mathcal{T}} \Upsilon(\bar{E}X(k) + \bar{F}(u(k) + \tilde{\mathcal{D}}(k)) - 1_N \otimes \bar{E_0}x_0(k)) = (1 - \bar{\lambda}\tau)\tilde{S}(k) - \bar{\vartheta}\tau sgn(\tilde{S}(k)). \quad (3.20)$$

Further, $u(k)$ is derived using Eq. (3.20) as

$$u(k) = [-(\sigma_s^{\mathcal{T}} \Upsilon \bar{F})^{-1}[\sigma_s^{\mathcal{T}} \Upsilon \bar{E}X(k) + \sigma_s^{\mathcal{T}} \Upsilon(-1_N \otimes \bar{E_0}x_0(k)) - (1 - \bar{\lambda}\tau)\tilde{S}(k) + \bar{\vartheta}\tau sgn(\tilde{S}(k))]] - \tilde{\mathcal{D}}(k). \quad (3.21)$$

This completes the proof.

Using protocol in Eq. (3.12), the global consensus error defined in Eq. (3.9) crosses the sliding surface, but no longer remain onto the surface rather move in a zig-zag style within a specified band called Quasi-Sliding Mode Band (QSMB) ([6]). To obtain the quasi-sliding mode band, let us consider reaching law defined in Eq. (3.18)

$$\tilde{S}_i(k+1) = (1 - \lambda_i\tau_i)\tilde{S}_i(k) - \vartheta_i\tau_i sgn(\tilde{S}_i(k)), \quad (3.22)$$

where τ_i is sampling period, $\lambda_i, \vartheta_i > 0$ are design parameters for the controller.

The sign of the first right-side term of the above equation is the same as $\tilde{S}_i(k)$ and the second term has the opposite sign of $\tilde{S}_i(k)$. Hence, according to the definition of QSMB defined in ([6]), the sgn of $\tilde{S}_i(k+1)$ must be opposite to that of $\tilde{S}_i(k)$. Thus, the above Eq. (3.22) can be written as

$$-\tilde{S}_i(k) = (1 - \lambda_i\tau_i)\tilde{S}_i(k) - \vartheta_i\tau_i sgn(\tilde{S}_i(k)). \quad (3.23)$$

Further evaluating Eq. (3.23)

$$2\tilde{S}_i(k) = \lambda_i\tau_i\tilde{S}_i(k) + \vartheta_i\tau_i sgn(\tilde{S}_i(k)). \quad (3.24)$$

Once the consensus error trajectory reaches inside the QSMB then $\tilde{S}_i(k)$ becomes $\tilde{S}_i(k) = \Delta_i$ and $sgn(\tilde{S}_i(k)) = 1$. Hence, Eq. (3.24) can be written as

$$\Delta_i = \frac{\vartheta_i\tau_i}{2 - \lambda_i\tau_i}. \quad (3.25)$$

The next theorem presents the analysis of fixed time steps required for the global consensus of homogeneous DMAS using the protocol defined in Eq. (3.12).

Theorem 3.1 *The number of time steps required to achieve the global consensus of DMAS (3.3) with DSM protocol (3.12) is $Max(k^\star + 1)$ where $k^\star$ for i^{th} agent is given by*

$$[k^\star] = \log_{(1-\lambda_i\tau_i)}\left(\frac{\Delta_i\lambda_i + \vartheta_i}{\vartheta_i + \lambda_i \mid \tilde{S}_i(0) \mid}\right), \quad (3.26)$$

where the notation $[k^\star]$ denotes the maximum integer below the actual real number $k^\star$.

Proof Based on the initial state of ith agent, there can be two cases: $\tilde{S}_i(0) > 0$ and $\tilde{S}_i(0) < 0$.

Case 1 : When $\tilde{s}_i(0) > 0$

From reaching law defined in Eq. (3.22), we may write following for ith agent:

For $k = 0$,

$$\tilde{S}_i(1) = (1 - \lambda_i \tau_i)\tilde{S}_i(0) - \vartheta_i \tau_i$$

and for $k = 1$,

$$\tilde{S}_i(2) = (1 - \lambda_i \tau_i)\tilde{S}_i(1) - \vartheta_i \tau_i.$$

Substituting $\tilde{S}_i(1)$ into $\tilde{S}_i(2)$, we may write

$$\tilde{S}_i(2) = (1 - \lambda_i \tau_i)[(1 - \lambda_i \tau_i)\tilde{S}_i(0) - \vartheta_i \tau_i] - \vartheta_i \tau_i$$

$$\vdots$$

Similarly for $k = z$, where z is positive integer, we may write

$$\tilde{S}_i(z+1) = (1 - \lambda_i \tau_i)^{z+1}\tilde{s}_i(0) - (1 - \lambda_i \tau_i)^{z}\vartheta_i \tau_i - (1 - \lambda_i \tau_i)\vartheta_i \tau_i - (1 - \lambda_i \tau_i)^{0}\vartheta_i \tau_i. \quad (3.27)$$

In general,

$$\tilde{S}_i(z+1) = (1 - \lambda_i \tau_i)^{z+1}\tilde{s}_i(0) - \vartheta_i \tau_i \times \sum_{\mathcal{P}=0}^{z}(1 - \lambda_i \tau_i)^{(z-\mathcal{P})} \quad (3.28)$$

$$\tilde{S}_i(z+1) = (1 - \lambda_i \tau_i)^{z+1}\tilde{s}_i(0) - h_i, \quad (3.29)$$

where

$$h_i = \vartheta_i \tau_i \times \sum_{\mathcal{P}=0}^{m}(1 - \lambda_i \tau_i)^{(z-\mathcal{P})}. \quad (3.30)$$

Comparing Eq. (3.30) with geometric progression series defined as $a\frac{(1-r^d)}{1-r}$, where a is the first term of the series, r is the common ratio, and d is the number of terms in the series, we get

$$a = 1, r = (1 - \lambda_i \tau_i), d = z + 1.$$

Thus (3.30) can be written as

$$h_i = \vartheta_i \tau_i \times \frac{1 - (1 - \lambda_i \tau_i)^{z+1}}{\lambda_i \tau_i}. \quad (3.31)$$

Substituting Eq. (3.31) into (3.29),

$$\tilde{S}_i(z+1) = (1-\lambda_i\tau_i)^{z+1}\tilde{S}_i(0) - \vartheta_i\tau_i \times \frac{1-(1-\lambda_i\tau_i)^{z+1}}{\lambda_i\tau_i}. \tag{3.32}$$

Now defining $k^\star = z+1$, we may write

$$\tilde{S}_i(k^*) = (1-\lambda_i\tau_i)^{k^\star}\tilde{S}_i(0) - \vartheta_i\tau_i \times \frac{1-(1-\lambda_i\tau_i)^{k^\star}}{\lambda_i\tau_i}. \tag{3.33}$$

From Eq. (3.33), we may infer that for $k \geq k^\star$ steps the local neighborhood error trajectory of ith agent first time reaches to the QSMB and stays within a specified band (3.25).

Now for finding the $k^\star$, consider QSMB given in Eq. (3.25)

$$\Delta_i = (1-\lambda_i\tau_i)^{k^\star}\tilde{S}_i(0) - \vartheta_i\tau_i \times \frac{1-(1-\lambda_i\tau_i)^{k^\star}}{\lambda_i\tau_i}. \tag{3.34}$$

From Eq. (3.34),

$$(1-\lambda_i\tau_i)^{k^\star} = \frac{\Delta_i\lambda_i\tau_i + \vartheta_i\tau_i}{(\vartheta_i\tau_i) + \lambda_i\tau_i \mid \tilde{S}_i(0) \mid}. \tag{3.35}$$

Applying logarithm scale on both sides of Eq. (3.35), we get

$$k^\star = \log_{(1-\lambda_i\tau_i)} \frac{\Delta_i\lambda_i + \vartheta_i}{\vartheta_i + \lambda_i \mid \tilde{s}_i(0) \mid}. \tag{3.36}$$

Case 2 : $\tilde{S}_i(0) < 0$

In this case, $\tilde{S}_i(k+1)$ will be negative for all $k \geq k^* + 1$. Hence, we may write

$$\tilde{S}_i(k^\star+1) \geq -(1-\lambda_i\tau_i)^{k^\star+1}\tilde{S}_i(0) + \vartheta_i\tau_i \times \frac{1-(1-\lambda_i\tau_i)^{k^\star+1}}{\lambda_i\tau_i}. \tag{3.37}$$

The above two cases show that $k^\star$ is the least integer such that the local neighboring error trajectory of the ith agent is guaranteed to cross the sliding mode band defined in Eq. (3.25) and stay within it. Hence, the state consensus of ith agent with leader agent begins within $k^\star + 1$ steps. Here, based on the initial condition of each agent states, the maximum $k^\star + 1$ steps required for a particular agent to reach the leader are considered as a required number of steps for consensus.

This completes the proof.

The next theorem presents the global stability of the homogeneous DMAS with the proposed DSM protocol (3.12).

Theorem 3.2 *The global stability of leader-following consensus using the protocol (3.12) is guaranteed if the closed-loop error dynamics in (3.9) for global homogeneous DMAS in (3.3) drives toward the global sliding surface and maintains on it for any gain $\bar{\lambda}$, $\bar{\vartheta} > 0$, and $0 < 1 - \bar{\lambda}\tau < 1$, provided the following conditions are true:*

$$0 \le \bar{\alpha} < \tilde{S}^{\mathcal{T}}(k)\tilde{S}(k), \tag{3.38}$$

where $\bar{\alpha} = [(1 - \bar{\lambda}\tau)\tilde{S}(k) - \bar{\vartheta}\tau sgn(\tilde{S}(k))]^{\mathcal{T}} * [(1 - \bar{\lambda}\tau)\tilde{S}(k) - \bar{\vartheta}\tau sgn(\tilde{S}(k))]$.

Proof Let us consider the Lyapunov function

$$\bar{\mathcal{V}}(k) = \tilde{S}^{\mathcal{T}}(k)\tilde{S}(k). \tag{3.39}$$

Equation (3.39) can be written with forward derivative function as

$$\Delta\bar{\mathcal{V}}_s(k) = \bar{\mathcal{V}}(k+1) - \bar{\mathcal{V}}(k), \tag{3.40}$$

$$\Delta\bar{\mathcal{V}}_s(k) = \tilde{S}^{\mathcal{T}}(k+1)\tilde{S}(k+1) - \tilde{S}^{\mathcal{T}}(k)\tilde{S}(k). \tag{3.41}$$

For stability, it is required that $\Delta\bar{\mathcal{V}}_s(k) < 0$.

Substituting $\tilde{S}(k+1)$ using Eq. (3.16), we can get

$$\begin{aligned}\Delta\bar{\mathcal{V}}_s(k) = [\sigma_s^{\mathcal{T}}\bar{\Upsilon}(X(k+1) - 1_N \otimes x_0(k+1))]^{\mathcal{T}}[\sigma_s^{\mathcal{T}}\bar{\Upsilon}(X(k+1)\\ - 1_N \otimes x_0(k+1))] - \tilde{S}^{\mathcal{T}}(k)\tilde{S}(k).\end{aligned} \tag{3.42}$$

Substituting $X(k+1)$ and $x_0(k+1)$ from Eqs. (3.3) and (3.4),

$$\begin{aligned}\Delta\bar{\mathcal{V}}_s(k) = {} & [\sigma_s^{\mathcal{T}}\bar{\Upsilon}[(\bar{E}X(k) + \bar{F}(u(k) + \bar{\mathcal{D}}(k)) - 1_N \otimes \tilde{E}_0x_0(k))]]^{\mathcal{T}}\\ & [\sigma_s^{\mathcal{T}}\bar{\Upsilon}[(\bar{E}X(k) + \bar{F}(u(k) + \bar{\mathcal{D}}(k)) - 1_N \otimes \tilde{E}_0x_0(k))]]\\ & -\tilde{S}^{\mathcal{T}}(k)\tilde{S}(k).\end{aligned} \tag{3.43}$$

Substituting the protocol defined in Eq. (3.12) into Eq. (3.43), we may get

$$\begin{aligned}\Delta\bar{\mathcal{V}}_s(k) = [(1 - \bar{\lambda}\tau)\tilde{S}(k) - \bar{\vartheta}\tau sgn(\tilde{S}(k))]^{\mathcal{T}} * [(1 - \bar{\lambda}\tau)\tilde{S}(k)\\ - \bar{\vartheta}\tau sgn(\tilde{S}(k))] - \tilde{S}^{\mathcal{T}}(k)\tilde{S}(k),\end{aligned} \tag{3.44}$$

denoting $\bar{\alpha} = [(1 - \bar{\lambda}\tau)\tilde{S}(k) - \bar{\vartheta}\tau sgn(\tilde{S}(k))]^{\mathcal{T}} * [(1 - \bar{\lambda}\tau)\tilde{S}(k) - \bar{\vartheta}\tau sgn(\tilde{S}(k))]$.

$$\Delta\bar{\mathcal{V}}_s(k) = \bar{\alpha} - \tilde{S}^{\mathcal{T}}(k)\tilde{S}(k). \tag{3.45}$$

From Eq. (3.45), the global stability of homogeneous DMAS can be achieved by tuning $\bar{\alpha}$ close to zero with the proper selection of $\bar{\lambda}$ and $\bar{\vartheta}$ such that $\Delta\bar{\mathcal{V}}_s(k) < 0$.

The main detriment of the global consensus protocol defined in Eq. (3.12) is the constant zig-zag motion during the sliding phase which demands constant control effort for the consensus which consumes more energy. To overcome this drawback, the Power rate reaching law [7] approach is proposed in the literature. In the Power rate reaching law approach, the control effort is more when the error trajectory is away from the surface and reduce when it is close to the surface.

3.3 Analysis for Number of Steps Required for Global Consensus of DMAS Using the Protocol with the Power Rate Reaching Law

In this section, we propose analysis for number of steps required for the global consensus using DSM protocol with Power rate reaching law.

Theorem 3.3 *For the global DMAS (3.3) is said to achieve the global consensus with the leader agent in a fixed-time steps using protocol given as*

$$u(k) = [-(\sigma_s^{\mathcal{T}} \bar{\Upsilon} \bar{F})^{-1}[\sigma_s^{\mathcal{T}} \bar{\Upsilon} \bar{E} X(k) + \sigma_s^{\mathcal{T}} \bar{\Upsilon}(-1_N \otimes \tilde{E}_0 x_0(k))(1 - \bar{\lambda}\tau)\tilde{S}(k) + \bar{\vartheta}\tau \mid \tilde{S}(k) \mid^{\eta} sgn(\tilde{S}(k))]] - \bar{\mathcal{D}}(k), \tag{3.46}$$

where $(\sigma_s^{\mathcal{T}} \bar{\Upsilon} \bar{F})^{-1}$ *is non-singular.*

Proof First, defining the sliding surface of the ith agent as

$$\tilde{S}_i(k) = \sigma_{s_i}^{\mathcal{T}} \bar{\delta}_i(k), \tag{3.47}$$

where $\sigma_{s_i}^{\mathcal{T}}$ is the sliding mode gain to be obtained using the pole-placement approach [5] and τ is the sampling period of the discrete-time system.

From Eq. (3.47), the global sliding surface can be defined as

$$\tilde{S}(k) = \sigma_s^{\mathcal{T}} \bar{\Lambda}(k). \tag{3.48}$$

Using Eq. (3.9), we may write

$$\tilde{S}(k) = \sigma_s^{\mathcal{T}} \bar{\Upsilon}(X(k) - 1_N \otimes x_0(k)). \tag{3.49}$$

Further,

$$\tilde{S}(k+1) = \sigma_s^{\mathcal{T}} \bar{\Upsilon}(X(k+1) - 1_N \otimes x_0(k+1)). \tag{3.50}$$

Substituting Eqs. (3.3) and (3.4) into Eq. (3.50), we may get

$$\tilde{S}(k+1) = \sigma_s^{\mathcal{T}} \bar{\Upsilon}(\bar{E} X(k) + \bar{F}(u(k) + \bar{\mathcal{D}}(k)) - 1_N \otimes \tilde{E}_0 x_0(k)). \tag{3.51}$$

Now, let us define the Power rate reaching law [7], for the ith agent

$$\tilde{S}_i(k+1) = (1-\lambda_i\tau_i)\tilde{S}_i(k) - \vartheta_i\tau_i \mid \tilde{S}_i(k) \mid^{\eta_i} sgn(\tilde{S}_i(k)). \tag{3.52}$$

From Eq. (3.52), we may derive the Power rate reaching law for the global homogeneous DMAS system as

$$\tilde{S}(k+1) = (1-\bar{\lambda}\tau)\tilde{S}(k) - \bar{\vartheta}\tau \mid \tilde{S}(k) \mid^{\eta} sgn(\tilde{S}(k)), \tag{3.53}$$

where $\bar{\vartheta} = [\vartheta_1, \vartheta_2, \vartheta_3, ...\vartheta_N] \in \mathbb{R}^{nN} > 0$, $\bar{\lambda} = [\lambda_1, \lambda_2, \lambda_3, ...\lambda_N] \in \mathbb{R}^{nN} > 0, 0 < (1-\bar{\lambda}\tau) < 1, \eta = [\eta_1, \eta_2, \eta_3...\eta_N]$.

Comparing Eqs. (3.51) and (3.53),

$$\begin{aligned}\sigma_s^{\mathcal{T}}\bar{\Upsilon}(\bar{E}X(k) + \bar{F}(u(k) + \bar{\mathcal{D}}(k)) - 1_N \otimes \tilde{E}_0 x_0(k)) &= (1-\bar{\lambda}\tau)\tilde{S}(k) \\ &- \bar{\vartheta}\tau \mid \tilde{S}(k) \mid^{\eta} sgn(\tilde{S}(k)).\end{aligned} \tag{3.54}$$

From Eq. (3.54), we may derive the DSM protocol for global consensus as

$$\begin{aligned}u(k) = [-(\sigma_s^{\mathcal{T}}\bar{\Upsilon}\bar{F})^{-1}[\sigma_s^{\mathcal{T}}\bar{\Upsilon}\bar{E}X(k) + \sigma_s^{\mathcal{T}}\bar{\Upsilon}(-1_N \otimes \tilde{E}_0 x_0(k)) - (1-\bar{\lambda}\tau)\tilde{S}(k) + \\ \bar{\vartheta}\tau \mid \tilde{S}(k) \mid^{\eta} sgn(\tilde{S}(k))]] - \bar{\mathcal{D}}(k).\end{aligned} \tag{3.55}$$

This completes the proof.

To derive the width of the Sliding Mode Band (SMB) using Power rate reaching law let us consider reaching law defined in Eq. (3.52)

$$\tilde{S}_i(k+1) = (1-\lambda_i\tau_i)\tilde{S}_i(k) - \vartheta_i\tau_i \mid \tilde{S}_i(k) \mid^{\eta_i} sgn(\tilde{S}_i(k)). \tag{3.56}$$

The sign of the first right-side term of the above equation is the same as $\tilde{S}_i(k)$, and the second term has the opposite sign of $\tilde{S}_i(k)$. Hence, according to the definition of QSMB, the sign of $\tilde{S}_i(k+1)$ must be opposite to that of $\tilde{S}_i(k)$. Thus, Eq. (3.56) can be written as

$$-\tilde{S}_i(k) = (1-\lambda_i\tau_i)\tilde{S}_i(k) - \vartheta_i\tau_i \mid \tilde{S}_i(k) \mid^{\eta_i} sgn(\tilde{S}_i(k)). \tag{3.57}$$

Further, Eq. (3.57) is written as

$$2\tilde{S}_i(k) = \lambda_i\tau_i\tilde{S}_i(k) + \vartheta_i\tau_i \mid \tilde{S}_i(k) \mid^{\eta_i} sgn(\tilde{S}_i(k)). \tag{3.58}$$

Once the consensus error (3.6) trajectory reaches inside the QSMB, then $\tilde{S}_i(k) = \Phi_i$, $sgn(\tilde{S}_i(k)) = 1$ for the ith agent and can be defined as

$$\Phi_i = \left(\frac{\vartheta_i\tau_i}{2-\lambda_i\tau_i}\right)^{\frac{1}{1-\eta_i}}. \tag{3.59}$$

From Eq. (3.59), we may infer that an appropriate selection of η_i reduces the chattering effect and increases the consensus convergence speed.

Theorem 3.4 *The number of time steps required to achieve the global consensus of DMAS (3.3) with DSM protocol (3.55) is $Max(k^\star + 1)$ where $k^\star$ for i^{th} agent is given by*

$$[k^\star] = \frac{[ln(|\tilde{S}_i(0)|^{1-\eta_i} + \frac{\vartheta_i}{\lambda_i}) - ln(\frac{\vartheta_i}{\lambda_i} + \Phi_i)]}{(1-\eta_i)\lambda_i\tau_i}, \tag{3.60}$$

where the notation $[k^\star]$ denotes the maximum integer bounded below the real number $k^\star$ and $\tilde{S}_i(0)$ is the initial value of the sliding mode surface.

Proof Consider the reaching law (3.56) for ith follower agent and assuming τ_i is small enough, we may use continuous-time derivative function

$$\dot{\tilde{S}}_i = \lim_{\delta \to 0} \frac{\tilde{S}_i(t+\delta(t)) - \tilde{S}_i(t)}{\delta(t)}. \tag{3.61}$$

Rewriting Eq. (3.61), we get

$$\dot{\tilde{S}}_i = (\tilde{S}_i(k+1) - \tilde{S}_i(k))/\tau_i = -\lambda_i \tilde{S}_i - \vartheta_i \tilde{S}_i^{\eta_i}. \tag{3.62}$$

Calculating reaching time for error trajectories,

$$\frac{d\tilde{S}_i}{dt} + \lambda_i \tilde{S}_i = -\vartheta_i \tilde{S}_i^{\eta_i}. \tag{3.63}$$

Multiplying both sides with $\tilde{s}_i^{-\eta_i}$,

$$\tilde{S}_i^{-\eta_i}\frac{d\tilde{S}_i}{dt} + \lambda_i \tilde{S}_i^{1-\eta_i} = -\vartheta_i. \tag{3.64}$$

Considering $z = \tilde{s}_i^{1-\eta_i}$ and taking derivative of it, we may get

$$\frac{dz}{dt} = (1-\eta_i)\tilde{S}_i^{-\eta_i}\frac{d\tilde{S}_i}{dt}. \tag{3.65}$$

Multiplying Eq. (3.64) by $(1-\eta_i)$ on both sides, we get

$$(1-\eta_i)\tilde{S}_i^{-\eta}\frac{d\tilde{S}_i}{dt} + (1-\eta_i)\lambda_i \tilde{S}_i^{1-\eta_i} = -(1-\eta_i)\vartheta_i. \tag{3.66}$$

Substituting Eq. (3.65) into Eq. (3.66), we get

$$\frac{dz}{dt} + (1-\eta_i)\lambda_i z = -(1-\eta_i)\vartheta_i. \tag{3.67}$$

Now from the first-order non-homogeneous differential equation defined as

$$\frac{dz}{dt} + \tilde{p}(t)z = \tilde{Q}(t), \tag{3.68}$$

whose general solution is given as ([8])

$$z = ce^{-\int \tilde{p}(t)dt} + e^{-\int \tilde{p}(t)dt} \int \tilde{Q}(t)e^{-\int \tilde{p}(t)dt} dt, \tag{3.69}$$

$$z = (ce^{-\int (1-\eta_i)\lambda_i dt} + e^{-\int (1-\eta_i)\lambda_i dt}) \times (\int [-(1-\eta_i)\vartheta_i] e^{-\int (1-\eta_i)\lambda_i dt} dt). \tag{3.70}$$

Combining with Eq. (3.70), the solution of Eq. (3.68) is given by

$$\begin{aligned} z &= ce^{-(1-\eta_i)\lambda_i t} + e^{-(1-\eta_i)\lambda_i t} \frac{([-(1-\eta_i)\vartheta_i]e^{(1-\eta_i)\lambda_i t})}{(1-\eta_i)\lambda_i}, \\ &= ce^{-(1-\eta_i)\lambda_i t} - \frac{\vartheta_i}{\lambda_i}. \end{aligned} \tag{3.71}$$

Considering $z(t) = \tilde{S}_i^{1-\eta_i}(t)$, we may write

$$\tilde{S}_i^{1-\eta_i}(t) = ce^{-(1-\eta_i)\lambda_i t} - \frac{\vartheta_i}{\lambda_i}. \tag{3.72}$$

When $t = 0$, we may find the solution of c from Eq. (3.72) as

$$c = |\tilde{S}_i(0)|^{(1-\eta_i)} + \frac{\vartheta_i}{\lambda_i}. \tag{3.73}$$

Substituting Eq. (3.73) into Eq. (3.72), we get

$$\tilde{S}_i^{1-\eta_i}(t) = \left[|\tilde{S}_i(0)|^{(1-\eta_i)} + \frac{\vartheta_i}{\lambda_i} \right] e^{-(1-\eta_i)\lambda_i t} - \frac{\vartheta_i}{\lambda_i}. \tag{3.74}$$

From Eq. (3.74), the reaching time from any initial condition to sliding mode band (SMB) for ith agent is defined as

$$\Phi_i = \left[|\tilde{S}_i(0)|^{(1-\eta_i)} + \frac{\vartheta_i}{\lambda_i} \right] e^{-(1-\eta_i)\lambda_i t} - \frac{\vartheta_i}{\lambda_i}. \tag{3.75}$$

Further, using logarithmic scale function on both sides of Eq. (3.75), we may get

$$t = \frac{\{ln[|\ \tilde{S}_i(0)\ |^{(1-\eta_i)} + \frac{\vartheta_i}{\lambda_i}] - ln[\Phi_i + \frac{\vartheta_i}{\lambda_i}]\}}{(1-\eta_i)\lambda_i}. \tag{3.76}$$

Let's take $\bar{\mathcal{X}} = ln[| \tilde{S}_i(0) |^{(1-\eta_i)} + \frac{\vartheta_i}{\lambda_i}]$, $\bar{\mathcal{Y}} = ln[\Phi_i + \frac{\vartheta_i}{\lambda_i}]$, and Eq. (3.76), and the discrete steps are calculated from any initial condition to SMB $\tilde{S}_i(k) = \Phi_i$ given by

$$k^{\star} = t/\tau_i = \frac{\bar{\mathcal{X}} - \bar{\mathcal{Y}}}{(1-\eta_i)\lambda_i\tau_i}. \tag{3.77}$$

From Eq. (3.77), it is noted that the parameters λ_i, ϑ_i, and η_i affect the convergence speed for the consensus.

This completes the proof.

The global stability of homogeneous DMAS using the DSM protocol (3.55) can be derived in similar form as given in Theorem 3.2.

3.4 Design, Validation, and Comparison of the Proposed Protocols on Homogeneous DMAS Comprise of 2-DOF Helicopter Systems

3.4.1 System Description and Modeling

Because of its broad nonlinear features, extremely cross-coupling impacts, and open-loop instability, the 2-DOF helicopter is a significant model from the control system engineering point of perspective. Figure 3.1 shows the 2-DOF helicopter model (fixed base) with two DC-motor-driven propellers. The front propeller controls the elevation of the nose over the pitch axis and the back propeller controls the rotational

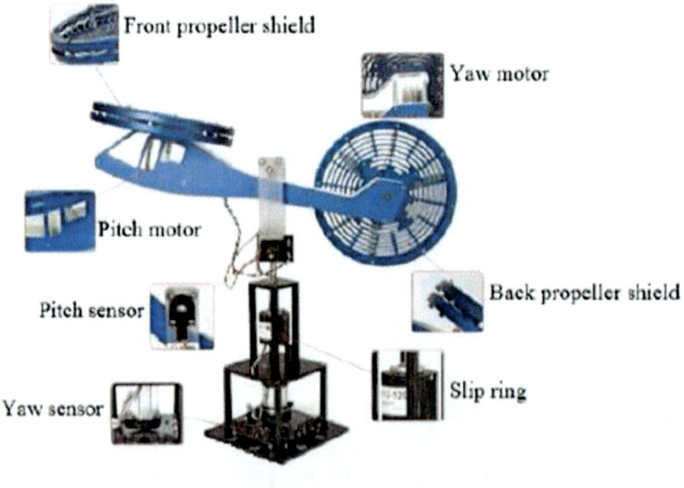

Fig. 3.1 2-DOF helicopter system

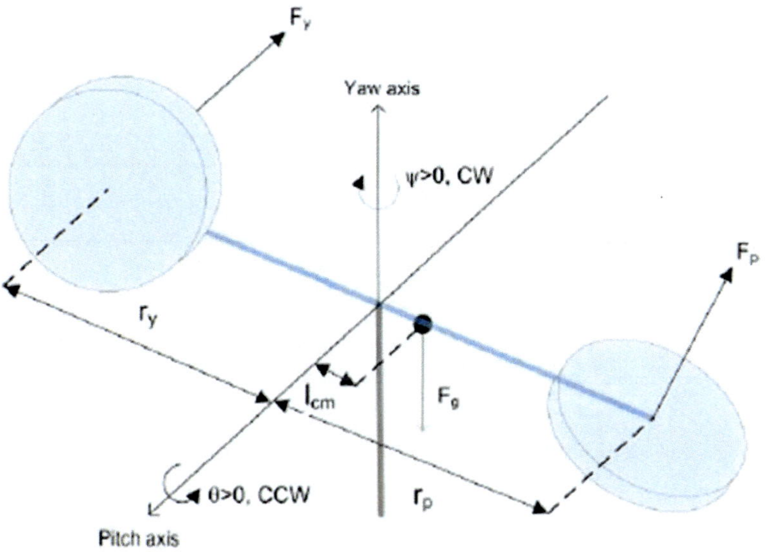

Fig. 3.2 Free body diagram of 2-DOF helicopter system

motion around the yaw axis. The voltages are ±24 V and ±15 V, respectively, across the pitch and yaw motors [9].

The 2-DOF helicopter model dynamics are shown in Fig. 3.2. The thrust forces F_p and F_y are applied across the pitch and yaw axes, respectively. The torques act at a distance r_p and r_y from the respective axis. The gravitational force F_g pulls down the helicopter nose. The center of mass acts at l distance from the pitch axis along the helicopter body length, where l is the distance between the center of mass and the intersection of the pitch and yaw axes. The notations used for the modeling of the 2-DOF helicopter system are given in Table 3.1.

The 2-DOF helicopter modeling conventions are as follows:

- The helicopter is horizontal when the pitch angle equals $\theta = 0$.
- The pitch angle increases positively, i.e, $\dot{\theta} > 0$, when the nose is moved upward and the body rotates in the counter-clockwise (CCW) direction.
- The yaw angle increases positively, i.e., $\dot{\psi} > 0$ when the body rotates in the clockwise (CW) direction.
- Pitch increases, $\theta > 0$, when the pitch thrust force is positive $F_p > 0$.
- Yaw increases, $\psi > 0$, when the yaw thrust force is positive, $F_y > 0$.

For pitch axis angle θ, the dynamic equation is written as

$$\ddot{\theta} = \frac{1}{J_{eq,p} + m_h l^2}[\{K_{pp}V_p + K_{py}V_y\} - \{B_p\dot{\theta} + m_h g l cos(\theta) \\ + m_h l^2 sin(\theta)cos(\theta)\dot{\psi}^2\}]. \tag{3.78}$$

Table 3.1 Notations used for modeling of 2-DOF helicopter system

θ	Pitch angle, deg
ψ	Yaw angle, deg
K_{pp}	Thrust force constant of yaw motor/propeller, N.m/V
K_{yy}	Thrust torque constant of yaw axis from yaw motor/propeller, N.m/V
K_{py}	Thrust torque constant acting on pitch axis from yaw motor/propeller, N.m/V
K_{yp}	Thrust torque constant acting on yaw axis from pitch motor/propeller, N.m/V
B_p	Viscous damping about pitch axis, N/V
B_y	Viscous damping about yaw axis, N/V
m_h	Total moving mass of the helicopter (body, two propellers, assemblies, etc.), kg
l	Center of mass length along helicopter body from pitch axis, meter(m)
J_{eq}, p	Total moment of inertia about pitch axis, kg.m^2
J_{eq}, y	Total moment of inertia about yaw axis, kg.m^2
V_p, V_y	Voltages applied to the front and rear motors
g	Gravitational constant, 9.8m/s^2

Considering $J_{tp} = J_{eq,p} + m_h l^2$, we may write Eq. (3.78) as

$$\ddot{\theta} = \frac{1}{J_{tp}}[\{K_{pp}V_p + K_{py}V_y\} - \{B_p\dot{\theta} + m_h g l cos(\theta) \\ + m_h l^2 sin(\theta)cos(\theta)\dot{\psi}^2\}]. \tag{3.79}$$

Similarly, for the yaw axis it is defined as

$$\ddot{\psi} = \frac{1}{J_{eq,y} + m_h l^2 cos^2(\theta)}[\{K_{yp}V_p + K_{yy}V_y\} - \{B_y\dot{\psi} \\ + 2m_h l^2 sin(\theta)cos(\theta)\dot{\psi}\dot{\psi}^2\}]. \tag{3.80}$$

Considering $J_{ty} = J_{eq,y} + m_h l^2 cos^2(\theta)$, we may write Eq. (3.80) as

$$\ddot{\psi} = \frac{1}{J_{ty}}[\{K_{yp}V_p + K_{yy}V_y\} - \{B_y\dot{\psi} + 2m_h l^2 sin(\theta)cos(\theta)\dot{\psi}\dot{\psi}^2\}]. \tag{3.81}$$

The state space model for ith system is defined as

$$x_i(k+1) = \tilde{E}x_i(k) + \tilde{F}(u_i(k) + \tilde{\mathcal{D}}_i(k)) \ \forall_i \in N, \tag{3.82}$$

where $\tilde{E} = \begin{bmatrix} 0\,0 & 1 & 0 \\ 0\,0 & 0 & 1 \\ 0\,0 & -\frac{B_p}{J_{Tp}} & 0 \\ 0\,0 & 0 & -\frac{B_y}{J_{Ty}} \end{bmatrix} \in \mathbb{R}^{n\times n}$ and $\tilde{F} = \begin{bmatrix} 0 & 0 \\ 0 & 0 \\ \frac{k_{pp}}{J_{Tp}} & \frac{k_{py}}{J_{Tp}} \\ \frac{k_{yp}}{J_{Ty}} & \frac{k_{yy}}{J_{Ty}} \end{bmatrix} \in \mathbb{R}^{n\times m}$ are the system matrix input matrix, respectively, and $x_i(k) = [\theta_i(k)\psi_i(k)\dot{\theta}_i(k)\dot{\psi}_i(k)] \in \mathbb{R}^n$, $u_i(k) \in \mathbb{R}^m$, and $\tilde{\mathcal{D}}_i \in \mathbb{R}^m$ are state vector, input vector, and matched disturbance acting on ith follower.

3.4.2 Simulation Results

Let us consider a homogeneous DMAS comprising four numbers of 2-DOF helicopter systems as shown in Fig. 3.3 for the validation of the DSM protocols proposed in this chapter. In the homogeneous DMAS configuration, one of the helicopter systems acts as a leader and the other three act as follower agents. The objective is to achieve the consensus of pitch angle and its velocity and yaw angle and its velocity of all the helicopter systems. Pitch position (θ) and its velocity ($\dot{\theta}$), yaw position(ψ) and its velocity ($\dot{\psi}$) states of 2-DOF helicopter system are to be considered as consensus parameters. As mentioned, the node of the leader is thought to have the same dynamics as the node of the follower agents. MATLAB R16 is used to perform the simulation. Discretizing the continuous-time system (3.82) at sampling period $\tau = 0.03$,

$$x_i(k+1) = \tilde{E}x_i(k) + \tilde{F}(u_i(k) + \tilde{\mathcal{D}}_i(k)) \;\; \forall_i \in N, \tag{3.83}$$

where

$$\tilde{E} = \begin{bmatrix} 1 & 0 & 0.0262 & 0 \\ 0 & 1 & 0 & 0.0285 \\ 0 & 0 & 0.7571 & 0 \\ 0 & 0 & 0 & 0.9004 \end{bmatrix} \tilde{F} = \begin{bmatrix} 0.0010 & 0 \\ 0.0001 & 0.0003 \\ 0.0620 & 0.0021 \\ 0.0069 & 0.0225 \end{bmatrix}.$$

Consider a MAS with three follower nodes and one leader node having communication graph topology as shown in Fig. 3.3. The notation of the leader node is given as 0, while the notation of the following nodes is given as 1, 2, 3, respectively. Each follower and leader agent's initial state is generated in the span$[-1, 1]$. Diagonal matrix $\tilde{D}$, adjacency matrix $\mathcal{A}$, pinning gain matrix $\tilde{\mathcal{B}}$, and Laplacian matrix $\tilde{L}$ are defined as

$$\tilde{D} = \begin{bmatrix} 2 & 0 & 0 \\ 0 & 2 & 0 \\ 0 & 0 & 2 \end{bmatrix}, \mathcal{A} = \begin{bmatrix} 0 & 1 & 1 \\ 1 & 0 & 1 \\ 1 & 1 & 0 \end{bmatrix}, \tilde{\mathcal{B}} = diag\{1, 1, 0\}, \tag{3.84}$$

$$\tilde{L} = \begin{bmatrix} 2 & -1 & -1 \\ -1 & 2 & -1 \\ -1 & -1 & 2 \end{bmatrix}. \tag{3.85}$$

The sliding mode gain for surface is calculated using pole-placement method as

$$\sigma_{s_i}^T = \begin{bmatrix} 93.64 & 52.39 & 13.17 & -0.0043 \\ 126.18 & 482.90 & -1.5626 & 45.98 \end{bmatrix}. \tag{3.86}$$

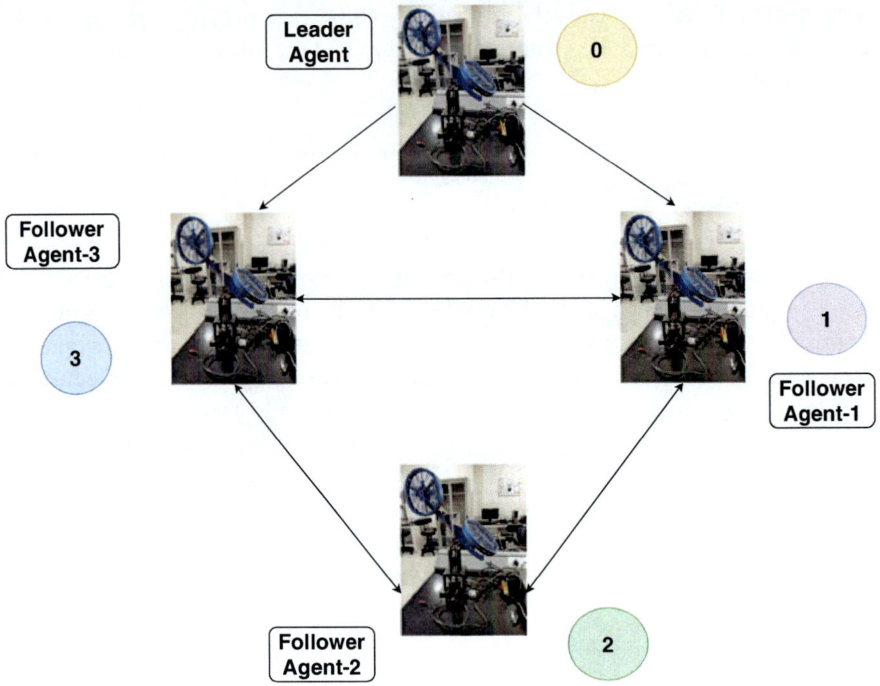

Fig. 3.3 Communication topology

A slow varying disturbance is applied to each DMAS with magnitude $\tilde{\mathcal{D}}_i(k) = 0.002 * cos(0.86k)$ to check the robustness of the protocol for the global consensus of homogeneous DMAS. The gain for each follower agent ϑ_i, λ_i is selected as 2 and 1, respectively, for the protocol using Gao's reaching law defined in Eq. (3.18). The value of η_i for ith agent is considered to be 0.6.

Figure 3.4 shows the simulation results of the follower agents pitch position and velocity states consensus with leader position and velocity states by applying proposed protocols using Gao's and Power rate reaching law. It is observed that the consensus of all the followers with a leader agent in fixed time steps is shown in Table 3.2.

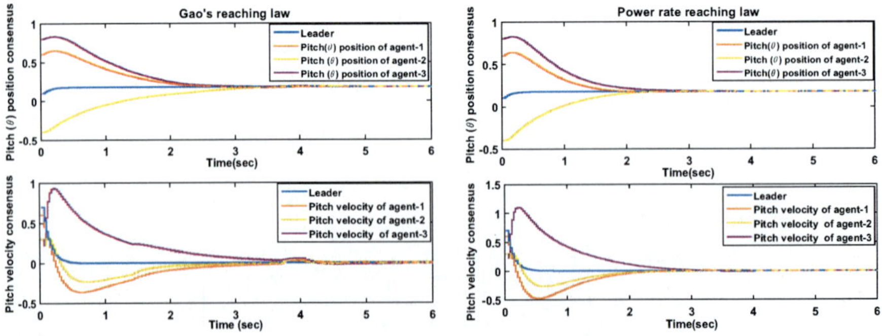

Fig. 3.4 Pitch position and velocity consensus

Table 3.2 Performance comparison of leader-following consensus

	DSM protocol using Gao's reaching law		DSM protocol using Power rate reaching law	
	Position consensus	Velocity consensus	Position consensus	Velocity consensus
Steps($k^\star$)	134	173	117	148
Order of chattering magnitude	$O(\tau)$	$O(\tau)$	$O(\tau^2)$	$O(\tau^2)$

Moreover, it is also observed from Table 3.2 that protocol using Power rate reaching law gives faster consensus convergence speed compared to Gao's reaching law.

Figure 3.5 shows the simulation results of yaw position and velocity states consensus of the follower with leader agent states and it is revealed that all the follower agents achieve the consensus with leader agent in fixed time steps shown in Table (3.2).

Figure 3.6 shows follower agent's consensus control effort (u), which has also been applied to individual agents in the same homogeneous DMAS network. The

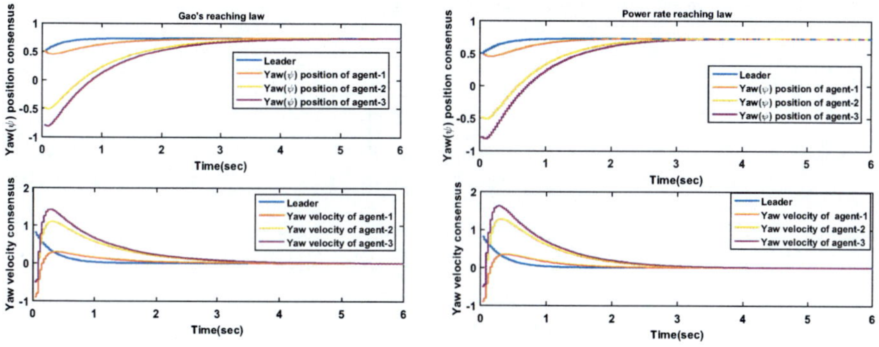

Fig. 3.5 Yaw position and velocity consensus

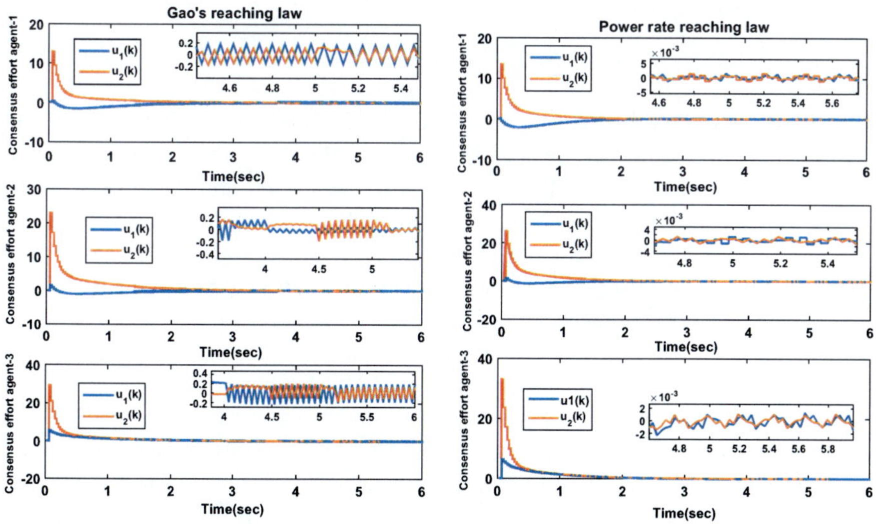

Fig. 3.6 Consensus effort (u) of individual agent

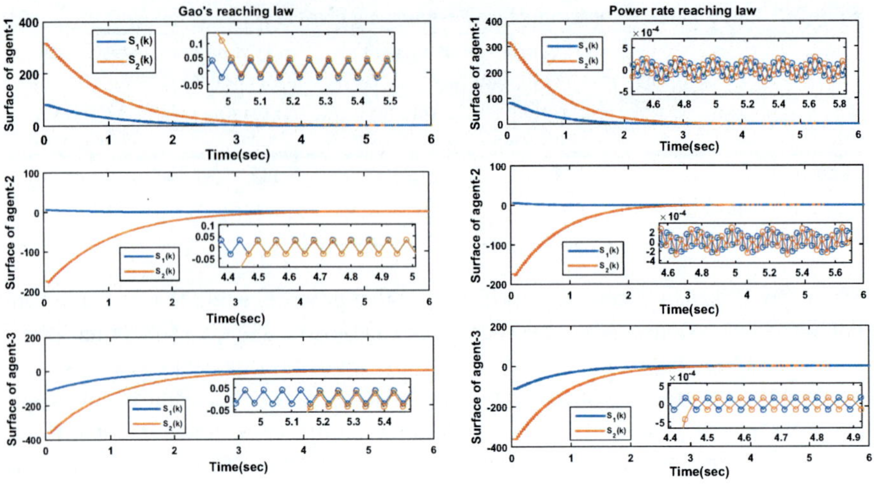

Fig. 3.7 Sliding variable of individual agent

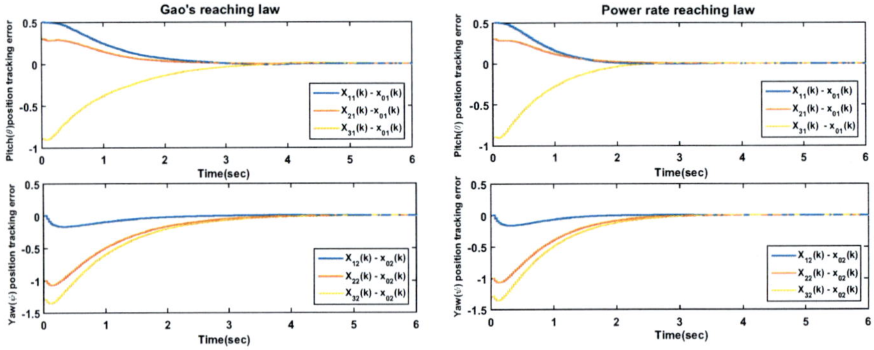

Fig. 3.8 Position consensus error

magnified window of the results for the individual follower agent shows the chattering effect.

Figure 3.7 shows the sliding surface (considered as leader-following consensus error) of the individual agent of homogeneous DMAS. It first crosses the sliding surface band called as QSMB from any initial condition and remains within this band. The sliding mode band changes from -0.03 to $+0.03$ ($O(\tau)$) determined using Eq. (3.25) for the case of Gao's reaching law. Likewise, sliding mode band varies in the range from -0.000251 to $+0.000251$ ($O(\tau^2)$) determined using Eq. (3.59) for the case of Power rate reaching law.

Figure 3.8 shows the error of pitch and yaw position of follower agents with the leader agent using two different proposed DSM protocols which subsequently converges to zero. Figure 3.9 shows the consensus error of pitch and yaw velocity with the leader agent using two distinct proposed DSM protocols using Gao's reaching law and Power rate reaching law. It can be noted that leader-following consensus error speedily reaches to zero in case of DSM protocol using power rate reaching law compared to DSM protocol using Gao's reaching law.

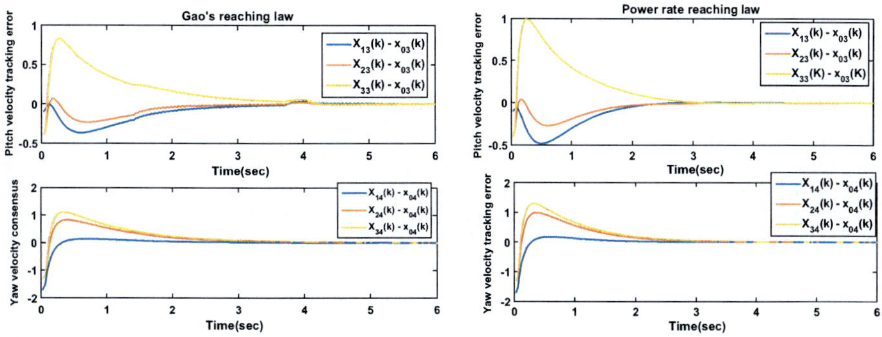

Fig. 3.9 Velocity consensus error

3.5 Conclusion

In this chapter, analytical study of discrete steps required for the global consensus of the follower agents with the leader using two distinct discrete-time sliding mode global consensus protocols based on Gao's reaching law and Power rate reaching law is discussed. A discrete multi-agent system which is a combination of leader and follower agents is configured with a fixed, undirected graph topology. It is inferred that the time required for global consensus due to Gao's reaching law takes more time in comparison to Power rate reaching law. The results reveal that Power rate reaching law has chattering magnitude $O(\tau^2)$ compared with Gao's reaching law as $O(\tau)$. The global stability condition is also derived for the DMAS using the Lyapunov function. Finally, both global consensus protocols are implemented in Simulink environment for a homogeneous multi-agent system comprising 2-DOF (degree of freedom) helicopter systems.

References

1. Patel, K., Mehta, A.: Discrete-time sliding mode control for leader following discrete-time multi-agent system. In: IECON 2018 - 44th Annual Conference of the IEEE Industrial Electronics Society, pp. 2288–2292 (2018)
2. Patel, K., Mehta, A.: Discrete higher order sliding mode protocol for leader-following consensus of heterogeneous discrete multi-agent system. In: Lecture Notes in Electrical Engineering, pp. 1–10. Springer, Singapore (2019)
3. Patel, K., Mehta, A.: Discrete-time event-triggered higher order sliding mode control for consensus of 2-dof robotic arms. European Journal of Control (2020)
4. Varga, R.S.: Geršgorin and His Circles. Springer, Berlin (2004)
5. Nguyen, T., Miao, Z., Pan, Y., Amini, N., Ugrinovskii, V., James, M.R.: Pole placement approach to coherent passive reservoir engineering for storing quantum information. Control Theory Technol. **15**, 193–205 (2017). Aug

6. Gao, W., Wang, Y., Homaifa, A.: Discrete-time variable structure control systems. IEEE Trans. Indust. Electron. **42**, 117–122 (1995)
7. Devika, K.B., Thomas, S.: Power rate exponential reaching law for enhanced performance of sliding mode control. Int. J. Control, Autom. Syst. **15**, 2636–2645 (2017)
8. Mondal, S.P., Roy, T.K.: Solution of first order linear non homogeneous ordinary differential equation in fuzzy environment based on lagrange multiplier method. J. Uncertainty Math. Sci. **2014**, 1–18 (2014)
9. Patel, K., Mehta, A.: Discrete-time sliding mode protocols for leader-following consensus of discrete multi-agent system with switching graph topology. Eur. J. Control **51**, 65–75 (2020)

Chapter 4
Discrete-Time Sliding Mode Protocols for Leader-Following Consensus of Discrete Multi-Agent System with Switching Graph Topology

4.1 Review of Switching Graph Theory

Remark 4.1 In continuation to brief discussion of switching graph topology in Sect. 2.2.2, let's denote a positive integer $\mathcal{S}$ and the digraph $\bar{G} = \{\bar{G}_1, \bar{G}_2, \bar{G}_3, ...\bar{G}_{\mathcal{S}}\}$ with vertices $V = \{0, 1, 2, ..N\}$ changes at switching signal $\Gamma(k)$ denoted by $\overline{G}_1 = \{V_1, E_1, A_1\}, \overline{G}_2 = \{V_2, E_2, A_2\}..., \overline{G}_{\mathcal{S}}=\{V_{\mathcal{S}}, E_{\mathcal{S}}, A_{\mathcal{S}}\}$. The union of these $\mathcal{S}$ digraphs is denoted by $\overline{G} = \bigcup_{i=1}^{\mathcal{S}} \overline{G}_i = \{\bigcup_{i=1}^{\mathcal{S}} V_i, \bigcup_{i=1}^{\mathcal{S}} E_i, \bigcup_{i=1}^{\mathcal{S}} A_i\}$, where $i = \{1, 2, 3, ...\mathcal{S}\}$.

Lemma 4.1 ([2]) *If the digraph* $\bar{G}$ *has a directed spanning tree, then the matrix* $(\tilde{L} + \tilde{\mathcal{B}})$ *is invertible.*

Lemma 4.2 ([3]) *Zero is a simple eigenvalue of L if and only if graph G has a rooted spanning tree.*

4.2 Introduction of Leader-Following Homogeneous DMAS

Consider the linear homogeneous DMAS comprises N agents having system dynamics of each agent as

$$x_i(k+1) = \tilde{E}x_i(k) + \tilde{F}(u_i(k) + \tilde{\mathcal{D}}_i(k)) \ \ \forall_i \in N, \tag{4.1}$$

where $i = 1, ..., N$, and $\tilde{E} \in \mathbb{R}^{n\times n}$ and $\tilde{F} \in \mathbb{R}^{n\times m}$ are the system matrix and input matrix of ith system, respectively. State vector $x_i(k) \in \mathbb{R}^n$ and the input vector $u_i(k) \in \mathbb{R}^m$, $\tilde{\mathcal{D}}_i \in \mathbb{R}^m$ is matched disturbance acting on ith system.

From Eq. (4.1), we may write the global homogeneous DMAS as

$$X(k+1) = (I_N \otimes \tilde{E})X(k) + (I_N \otimes \tilde{F})(u(k) + \tilde{\mathcal{D}}(k)), \tag{4.2}$$

K. Patel and A. Mehta, *Discrete-Time Sliding Mode Protocols for Discrete Multi-Agent System*, Studies in Systems, Decision and Control 303,
https://doi.org/10.1007/978-981-15-6311-9_4

where state vector $X(k) = [x_1(k), x_2(k)...x_N(k)]^T \in \mathbb{R}^{nN}$ and input vector $u(k) = [u_1(k), u_2(k)...u_N(k)]^T \in \mathbb{R}^{mN}$, disturbance vector $\tilde{\mathcal{D}}(k) = [\mathcal{D}_1(k), \mathcal{D}_2(k), \mathcal{D}_3(k), ... \mathcal{D}_N(k)]^T \in \mathbb{R}^{mN}$, $\otimes$ denotes the Kronecker product.

Rewriting Eq. (4.2) in more generic form as

$$X(k+1) = \bar{E}X(k) + \bar{F}(u(k) + \tilde{\mathcal{D}}(k)), \tag{4.3}$$

where $\bar{E} = (I_N \otimes \tilde{E})$ and $\bar{F} = (I_N \otimes \tilde{F})$.

Let us define the autonomous leader dynamics as

$$x_0(k+1) = \tilde{E}x_0(k), \tag{4.4}$$

where $x_0(k) \in \mathbb{R}^n$ is the state vector of the leader.

Problem statement: To design a robust DSM protocol for global consensus of homogeneous DMAS such that all the follower agents (4.3) achieve consensus with the leader (4.4) agent using the neighborhood agent information with switching graph topology.

It is assumed that the leader node can be observed from a small subset of nodes in graph $\bar{G}$. If ith agent is connected to the leader then particularly this edge is said to exist with weighting switching gain $a_{i0} > 0$. The agent with $a_{i0} > 0$ is referred as pinning node [4–6].

Let us define the error function for the leader-follower network

$$\delta_i(k) = \sum_{j \in N} a_{ij}[x_i(k) - x_j(k)] + a_{i0}[x_i(k) - x_0(k)]. \tag{4.5}$$

Using Eq. (4.5) and according to Lemmas 4.1 and 4.2, the consensus error in global form and according to switching graph topology can be rewritten as

$$\bar{\Lambda}(k) = ((I_N + \tilde{D}_{\bar{G}_{\Gamma[k]}} + \tilde{\mathcal{B}}_{\bar{G}_{\Gamma[k]}})^{-1})(\tilde{L}_{\bar{G}_{\Gamma[k]}} + \tilde{\mathcal{B}}_{\bar{G}_{\Gamma[k]}})) \otimes I_n)\tilde{x},$$

where $\tilde{x} = X(k) - 1_N \otimes x_0(k) \in \mathbb{R}^{nN}$. The eigenvalues of the weighted matrix $\tilde{\xi}_{\bar{G}_{\Gamma[k]}} = ((I_N + \tilde{D}_{\bar{G}_{\Gamma[k]}} + \tilde{\mathcal{B}}_{\bar{G}_{\Gamma[k]}})^{-1})(\tilde{L}_{\bar{G}_{\Gamma[k]}} + \tilde{\mathcal{B}}_{\bar{G}_{\Gamma[k]}}))$ are obtained using Gershgorin circle criteria [7], which lies inside the unit circle for a stable graph.

For ease of mathematical operations, let us express Eq. (4.2) as

$$\bar{\Lambda}(k) = (\tilde{\xi}_{\bar{G}_{\Gamma[k]}} \otimes I_n)\tilde{x}.$$

$$\bar{\Lambda}(k) = \bar{\Upsilon}_{\bar{G}_{\Gamma[k]}}(X(k) - 1_N \otimes x_0(k)).$$

Then

$$\bar{\Lambda}(k+1) = \bar{\Upsilon}_{\bar{G}_{\Gamma[k]}}(X(k+1) - 1_N \otimes x_0(k+1)).$$

$$\bar{\Lambda}(k+1) = \bar{\Upsilon}_{\bar{G}_{\Gamma[k]}}(\bar{E}X(k) + \bar{F}(u(k) + \tilde{\mathcal{D}}(k)) - 1_N \otimes \tilde{E}x_0(k)).$$

4.3 DSM Protocol for the Global Consensus of Homogeneous DMAS with Switching Graph Topology Using Gao's Reaching Law

In this section, we present DSM protocol for the global consensus of homogeneous DMAS using Gao's reaching law in the form of Theorem 4.1. The condition for global stability of the homogeneous DMAS with the proposed DSM protocol is also derived.

Theorem 4.1 *For the given homogeneous DMAS (4.3), consensus with the leader (4.4) is achieved using a DSM protocol given by*

$$u(k) = [-(\sigma_s^{\mathcal{T}} \bar{\Upsilon}_{\bar{G}_{\Gamma[k]}} \bar{F})^{-1}[\sigma_s^{\mathcal{T}} \bar{\Upsilon}_{\bar{G}_{\Gamma[k]}} \bar{E} X(k) + \sigma_s^{\mathcal{T}} \bar{\Upsilon}_{\bar{G}_{\Gamma[k]}} (-1_N \otimes \tilde{E}_0 x_0(k)) \\ -(1 - \bar{\lambda}\tau)\tilde{S}(k) + \bar{\vartheta}\tau sgn(\tilde{S}(k))]] - \bar{\mathcal{D}}(k), \tag{4.6}$$

where $(\sigma_s^{\mathcal{T}} \bar{\Upsilon}_{\bar{G}_{\Gamma[k]}} \bar{F})^{-1}$ *is non-singular and* $\sigma_s^{\mathcal{T}} = I_N \otimes \sigma_{s_i}^{\mathcal{T}}$.

Proof Let us define the sliding surface using Eq. (4.5) for an individual ith agent as

$$\tilde{S}_i(k) = \sigma_{s_i}^{\mathcal{T}} \bar{\delta}_i(k), \tag{4.7}$$

where $\sigma_{s_i}^{\mathcal{T}}$ is the sliding gain to be obtained using the pole-placement approach.

In the leader-follower network if ith agent is connected to the leader then the sliding surface can be written as

$$\tilde{S}(k) = \sigma_s^{\mathcal{T}} \bar{\Lambda}(k). \tag{4.8}$$

Using Eq. (4.2) and Lemma 4.2, we may write a sliding surface for the global system (4.3) as

$$\tilde{S}(k) = \sigma_s^{\mathcal{T}} \bar{\Upsilon}_{\bar{G}_{\Gamma[k]}} (X(k) - 1_N \otimes x_0(k)). \tag{4.9}$$

Further,

$$\tilde{S}(k+1) = \sigma_s^{\mathcal{T}} \bar{\Upsilon}_{\bar{G}_{\Gamma[k]}} (X(k+1) - 1_N \otimes x_0(k+1)). \tag{4.10}$$

Substituting Eqs. (4.3) and (4.4) into Eq. (4.10), we may get

$$\tilde{S}(k+1) = \sigma_s^{\mathcal{T}} \bar{\Upsilon}_{\bar{G}_{\Gamma[k]}} (\bar{E} X(k) + \bar{F}(u(k) + \bar{\mathcal{D}}(k)) - 1_N \otimes \tilde{E} x_0(k)). \tag{4.11}$$

Now motivated from the reaching law proposed in [8], let us define the consensus reaching law for ith agent as

$$\tilde{S}_i(k+1) = (1 - \lambda_i \tau_i)\tilde{S}_i(k) - \vartheta_i \tau_i sgn(\tilde{S}_i(k)). \tag{4.12}$$

Then global consensus reaching law for global homogeneous DMAS can be defined as

$$\tilde{S}(k+1) = (1 - \bar{\lambda}\tau)\tilde{S}(k) - \bar{\vartheta}\tau sgn(\tilde{S}(k)), \tag{4.13}$$

where $\bar{\vartheta} = [\vartheta_1, \vartheta_2, \vartheta_3, ...\vartheta_N] \in \mathbb{R}^{nN} > 0$, $\bar{\lambda} = [\lambda_1, \lambda_2, \lambda_3, ...\lambda_N] \in \mathbb{R}^{nN} > 0$, $0 < (1 - \bar{\lambda}\tau) < 1$, $\eta = [\eta_1, \eta_2, \eta_3 ... \eta_N]$.

Comparing Eqs. (4.11) and (4.13), we may write

$$\begin{aligned}\sigma_s^{\mathcal{T}}\bar{\Upsilon}_{\bar{G}_{\Gamma[k]}}(\bar{E}X(k) + \bar{F}(u(k) + \bar{\mathcal{D}}(k)) - 1_N \otimes \tilde{E}x_0(k)) = \\ (1 - \bar{\lambda}\tau)\tilde{S}(k) - \bar{\vartheta}\tau sgn(\tilde{S}(k)).\end{aligned} \tag{4.14}$$

Further, Eq. (4.14) can be expressed in terms of control protocol as

$$\begin{aligned}u(k) = [-(\sigma_s^{\mathcal{T}}\bar{\Upsilon}_{\bar{G}_{\Gamma[k]}}\bar{F})^{-1}[\sigma_s^{\mathcal{T}}\bar{\Upsilon}_{\bar{G}_{\Gamma[k]}}\bar{E}X(k) + \sigma_s^{\mathcal{T}}\bar{\Upsilon}_{\bar{G}_{\Gamma[k]}}(-1_N \otimes \tilde{E}_0 x_0(k)) - \\ (1 - \bar{\lambda}\tau)\tilde{S}(k) + \bar{\vartheta}\tau sgn(\tilde{S}(k))]] - \bar{\mathcal{D}}(k)\end{aligned} \tag{4.15}$$

This completes the proof.

As per protocol (4.15), the consensus error trajectory reaches to the surface in a zig-zag step within specified band called Quasi-Sliding Mode Band (QSMB). So the width of consensus error QSMB for ith agent (Δ_i) is defined using Eq. (4.12)

$$\Delta_i = \left(\frac{\vartheta_i \tau_i}{2 - \lambda_i \tau_i}\right). \tag{4.16}$$

Theorem 4.2 presents the global stability of the homogeneous DMAS with proposed DSM protocol (4.15).

The global stability of the homogeneous DMAS with the proposed DSM protocol (4.15) is obtained in similar way as Theorem 3.2. Only the difference in switching graph is that the proposed protocol (4.15) has ensured that $\Delta\mathcal{V}_s(k) < 0$ for non switching instants.

Theorem 4.2 *The overall global stability of the DMAS with switching graph topology using the proposed protocol defined in Eq. (4.15) is achieved as same using Theorem 3.2 with consideration of $\bar{\Upsilon}$ function of graph theory and it is written for the overall graph as* $\bar{\Upsilon} = \bigcup_{i=1}^{S} \bar{\Upsilon}^i_{\bar{G}_{\Gamma[k]}}$.

The main lacking point of the protocol defined in Eq. (4.15) is chattering and continuous consensus control effort required to enforce the consensus error trajectory to sliding surface during reaching and sliding phase which may induce serious problems during implementation. To overcome these drawbacks, we design the protocol motivated by Power rate reaching law [9] in which the consensus effort is more when the consensus error trajectory is away from the surface and reduces the effort when the consensus error trajectory is nearer to the surface.

4.4 DSM Protocol for the Consensus of Homogeneous DMAS Using Power Rate Reaching Law

In this section, the DSM protocol for the global consensus of homogeneous DMAS is designed using Power rate reaching law in the form of Theorem 4.3. The condition for global stability of the homogeneous DMAS with the proposed DSM protocol is also derived.

Theorem 4.3 *For the given homogeneous DMAS (4.3), the consensus with a leader (4.4) is achieved using a DSM protocol given by*

$$\begin{aligned} u(k) = [-(\sigma_s^{\mathcal{T}} \bar{\Upsilon}_{\bar{G}_{\Gamma[k]}} \bar{F})^{-1} [\sigma_s^{\mathcal{T}} \bar{\Upsilon}_{\bar{G}_{\Gamma[k]}} \bar{E} X(k) + \sigma_s^{\mathcal{T}} \bar{\Upsilon}_{\bar{G}_{\Gamma[k]}} (-1_N \otimes \tilde{E}_0 x_0(k)) \\ -(1 - \bar{\lambda}\tau)\tilde{S}(k) + \bar{\vartheta}\tau |\tilde{S}(k)|^{\eta} sgn(\tilde{S}(k))]] - \bar{\mathcal{D}}(k), \end{aligned} \tag{4.17}$$

where $(\sigma_s^{\mathcal{T}} \bar{\Upsilon}_{\bar{G}_{\Gamma[k]}} \bar{F})^{-1}$ *is non-singular and* $\sigma_s^{\mathcal{T}} = I_N \otimes \sigma_{s_i}^{\mathcal{T}}$.

Proof Let us define the sliding surface using Eq. (4.5) for an individual ith agent as

$$\tilde{S}_i(k) = \sigma_{s_i}^{\mathcal{T}} \bar{\delta}_i(k), \tag{4.18}$$

where $\sigma_{s_i}^{\mathcal{T}}$ is the sliding gain to be obtained using the pole-placement approach.

In the leader-follower network, if ith agent is connected to the leader then the sliding surface can be written as

$$\tilde{S}(k) = \sigma_s^{\mathcal{T}} \bar{\Lambda}(k). \tag{4.19}$$

Using Eq. (4.2), we may write a sliding surface for the global system (4.3) as

$$\tilde{S}(k) = \sigma_s^{\mathcal{T}} \bar{\Upsilon}_{\bar{G}_{\Gamma[k]}} (X(k) - 1_N \otimes x_0(k)). \tag{4.20}$$

Further,

$$\tilde{S}(k+1) = \sigma_s^{\mathcal{T}} \bar{\Upsilon}_{\bar{G}_{\Gamma[k]}} (X(k+1) - 1_N \otimes x_0(k+1)). \tag{4.21}$$

Substituting Eqs. (4.3) and (4.4) into Eq. (4.21), we may get

$$\tilde{S}(k+1) = \sigma_s^{\mathcal{T}} \bar{\Upsilon}_{\bar{G}_{\Gamma[k]}} (\bar{E} X(k) + \bar{F}(u(k) + \bar{\mathcal{D}}(k)) - 1_N \otimes \tilde{E} x_0(k)). \tag{4.22}$$

Now motivated from the reaching law proposed in [8], let us define the consensus reaching law for ith agent as

$$\tilde{S}_i(k+1) = (1-\lambda_i\tau_i)\tilde{S}_i(k) - \vartheta_i\tau_i|\tilde{S}_i(k)|^{\eta_i} sgn(\tilde{S}_i(k)), \tag{4.23}$$

where $\bar{\vartheta} = [\vartheta_1, \vartheta_2, \vartheta_3, ...\vartheta_N] \in \mathbb{R}^{nN} > 0$, $\bar{\lambda} = [\lambda_1, \lambda_2, \lambda_3, ...\lambda_N] \in \mathbb{R}^{nN} > 0$, $0 < (1-\bar{\lambda}\tau) < 1$, $\eta = [\eta_1, \eta_2, \eta_3 ... \eta_N]$.

Then the global consensus reaching law for global homogeneous DMAS can be defined as

$$\tilde{S}(k+1) = (1-\bar{\lambda}\tau)\tilde{S}(k) - \bar{\vartheta}\tau|\tilde{S}(k)|^{\eta} sgn(\tilde{S}(k)). \tag{4.24}$$

Comparing Eqs. (4.22) and (4.24), we may write

$$\begin{aligned} \sigma_s^{\mathcal{T}}\bar{\Upsilon}_{\bar{G}_{\Gamma[k]}}(\bar{E}X(k) + \bar{F}(u(k) + \bar{\mathcal{D}}(k)) - 1_N \otimes \tilde{E}x_0(k)) = \\ (1-\bar{\lambda}\tau)\tilde{S}(k) - \bar{\vartheta}\tau|\tilde{S}(k)|^{\eta} sgn(\tilde{S}(k)). \end{aligned} \tag{4.25}$$

Further, Eq. (4.25) can be expressed in terms of control protocol as

$$\begin{aligned} u(k) = [-(\sigma_s^{\mathcal{T}}\bar{\Upsilon}_{\bar{G}_{\Gamma[k]}}\bar{F})^{-1}[\sigma_s^{\mathcal{T}}\bar{\Upsilon}_{\bar{G}_{\Gamma[k]}}\bar{E}X(k) + \sigma_s^{\mathcal{T}}\bar{\Upsilon}_{\bar{G}_{\Gamma[k]}}(-1_N \otimes \tilde{E}_0 x_0(k)) \\ -(1-\bar{\lambda}\tau)\tilde{S}(k) + \bar{\vartheta}\tau|\tilde{S}(k)|^{\eta} sgn(\tilde{S}(k))]] - \bar{\mathcal{D}}(k). \end{aligned} \tag{4.26}$$

This completes the proof.

The consensus of leader and follower is obtained using DSM protocol based on the Power rate reaching law. The sliding mode band in which system remains steadily is defined for ith agent from Eq. (4.26) as

$$\Phi_i = \left(\frac{\vartheta_i\tau_i}{1-\lambda_i\tau_i}\right)^{\frac{1}{1-\eta_i}}. \tag{4.27}$$

Proper selection of η_i reduces the chattering effect and enhances the robustness.

The next theorem presents the global stability of the homogeneous DMAS with proposed DSM protocol (4.15).

Theorem 4.4 *The global stability of the homogeneous DMAS with the proposed DSM protocol using Power rate reaching law (4.15) is ensured in similar manner as defined in Theorem 3.2 and the Lyapunov function $\Delta\bar{V}_s(k) < 0$ for non switching interval. The term $\bar{\Upsilon}$ is a function of graph theory and changes according to different graph topology switched at distinct time interval therefore it is written for the overall graph as $\bar{\Upsilon} = \bigcup_{i=1}^{S} \bar{\Upsilon}^i_{\bar{G}_{\Gamma[k]}}$.*

4.5 Design, Validation, and Comparison of the Proposed Protocols on Homogeneous DMAS Comprise 2-DOF Helicopter Systems

4.6 Simulation Results

In this section, pitch position (θ) and its velocity ($\dot{\theta}$) and yaw position (ψ) and its velocity ($\dot{\psi}$) states of 2-DOF helicopter system defined in Sect. 3.4.1 are to be considered as a single agent parameter for consensus. As discussed, it is assumed here that the leader agent has same dynamics as the follower agents. The simulation is carried out using MATLAB R15.

Discretizing the continuous-time system at $\tau = 0.03$ sampling period, we obtain the discrete-time system model as

$$x_i(k+1) = \tilde{E}x_i(k) + \tilde{F}(u_i(k) + \tilde{\mathcal{D}}_i(k)) \ \forall_i \in N, \tag{4.28}$$

where $\tilde{E} = \begin{bmatrix} 1 & 0 & 0.0262 & 0 \\ 0 & 1 & 0 & 0.0285 \\ 0 & 0 & 0.7571 & 0 \\ 0 & 0 & 0 & 0.9004 \end{bmatrix}$, $\tilde{F} = \begin{bmatrix} 0.0010 & 0 \\ 0.0001 & 0.0003 \\ 0.0620 & 0.0021 \\ 0.0069 & 0.0225 \end{bmatrix}$.

Consider MAS with five follower nodes and one leader node as undirected switching graph topology. The leader node notation is given as 0 while follower node notation is given as $1, 2, 3, 4, 5$, respectively. Initial states of each follower agent are generated randomly in the span $[-1, 1]$, while the initial state of the leader node is $[0.1, 0.5, 0.7, 0.8]^T$ $\mathcal{A}$ pinning gains matrix $\mathcal{B}$ and Laplacian matrix of individual graph topology $\bar{G}_\Gamma[p]$ are defined as (Fig. 4.1)

$$\tilde{L}_1 = \begin{bmatrix} 1 & -1 & 0 & 0 & 0 \\ -1 & 2 & -1 & 0 & 0 \\ 0 & -1 & 1 & 0 & 0 \\ 0 & 0 & 0 & 0 & 0 \\ 0 & 0 & 0 & 0 & 0 \end{bmatrix}, \tilde{L}_2 = \begin{bmatrix} 0 & 0 & 0 & 0 & 0 \\ 0 & 0 & 0 & 0 & 0 \\ 0 & 0 & 1 & -1 & 0 \\ 0 & 0 & -1 & 2 & -1 \\ 0 & 0 & 0 & -1 & 1 \end{bmatrix}, \tilde{L}_3 = \begin{bmatrix} 0 & 0 & 0 & 0 & 0 \\ 0 & 1 & -1 & 0 & 0 \\ 0 & -1 & 2 & -1 & 0 \\ 0 & 0 & -1 & 1 & 0 \\ 0 & 0 & 0 & 0 & 0 \end{bmatrix}. \tag{4.29}$$

Further, pinning gain matrices are also defined as

$$\tilde{\mathcal{B}}_1 = \begin{bmatrix} 1 & 0 & 0 & 0 & 0 \\ 0 & 0 & 0 & 0 & 0 \\ 0 & 0 & 0 & 0 & 0 \\ 0 & 0 & 0 & 0 & 0 \\ 0 & 0 & 0 & 0 & 0 \end{bmatrix}, \tilde{\mathcal{B}}_2 = \begin{bmatrix} 0 & 0 & 0 & 0 & 0 \\ 0 & 0 & 0 & 0 & 0 \\ 0 & 0 & 0 & 0 & 0 \\ 0 & 0 & 0 & 0 & 0 \\ 0 & 0 & 0 & 0 & 1 \end{bmatrix}, \tilde{\mathcal{B}}_3 = \begin{bmatrix} 0 & 0 & 0 & 0 & 0 \\ 0 & 0 & 0 & 0 & 0 \\ 0 & 0 & 0 & 0 & 0 \\ 0 & 0 & 0 & 1 & 0 \\ 0 & 0 & 0 & 0 & 0 \end{bmatrix}. \tag{4.30}$$

The sliding gain for surface defined in Eq. (4.7) is calculated using the pole-placement method which comes out to be

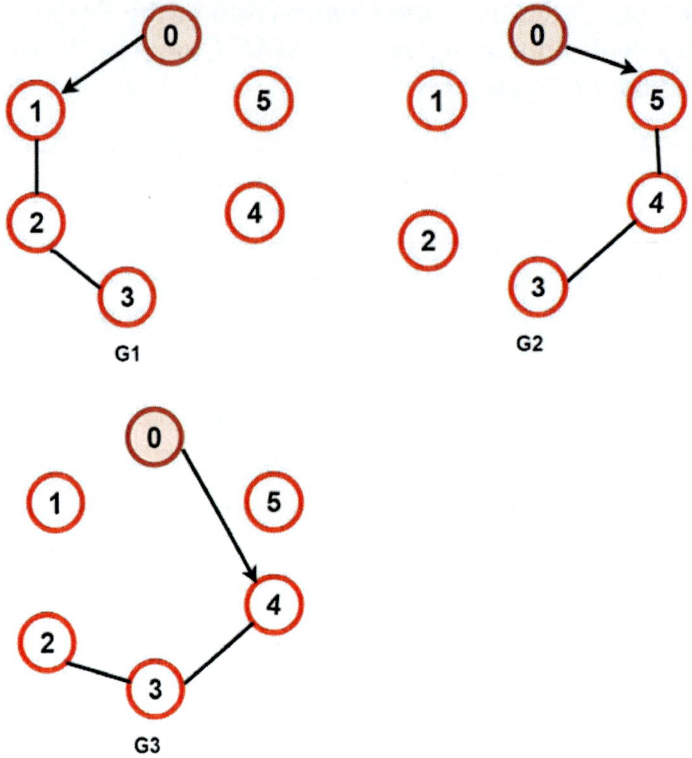

Fig. 4.1 Switching graph topology

$$\sigma_{s_i}^{\mathcal{T}} = \begin{bmatrix} 93.64 & 52.39 & 13.17 & -0.0043 \\ 126.18 & 482.90 & -1.5626 & 45.98 \end{bmatrix}. \tag{4.31}$$

In order to check the robustness of the derived protocol for the global system, a slow varying disturbance is applied to each DMAS with a magnitude $\tilde{\mathcal{D}}_i(k) = 0.002 * cos(0.86k)$ for individual graph topology. The gains for follower agents ϑ_i and λ_i are chosen as 2 and 1, respectively, in the case of Gao's reaching law and the gain value of η_i in the case of Power rate reaching law is taken as 0.6. Proper selection of the value η decreases the sliding mode band in Power rate reaching law.

Figure 4.2 shows the comparison results of the pitch position and velocity states consensus with leader position and velocity states using proposed DSM protocol with Gao's and Power rate reaching law using switching graph topology for the graph G_1, G_2, G_3. Graph G_1 has changed to G_2 after 50 number of steps. Simultaneously graph G_2 has changed to G_3 and remain stable for 49 number of steps and after that G_3 graph topology is steady for another 49 number of steps. In this way, we switch the topology and find that consensus begins after 148 number of steps. However, in this case, Power rate reaching law gives faster convergence speed compared to Gao's reaching law.

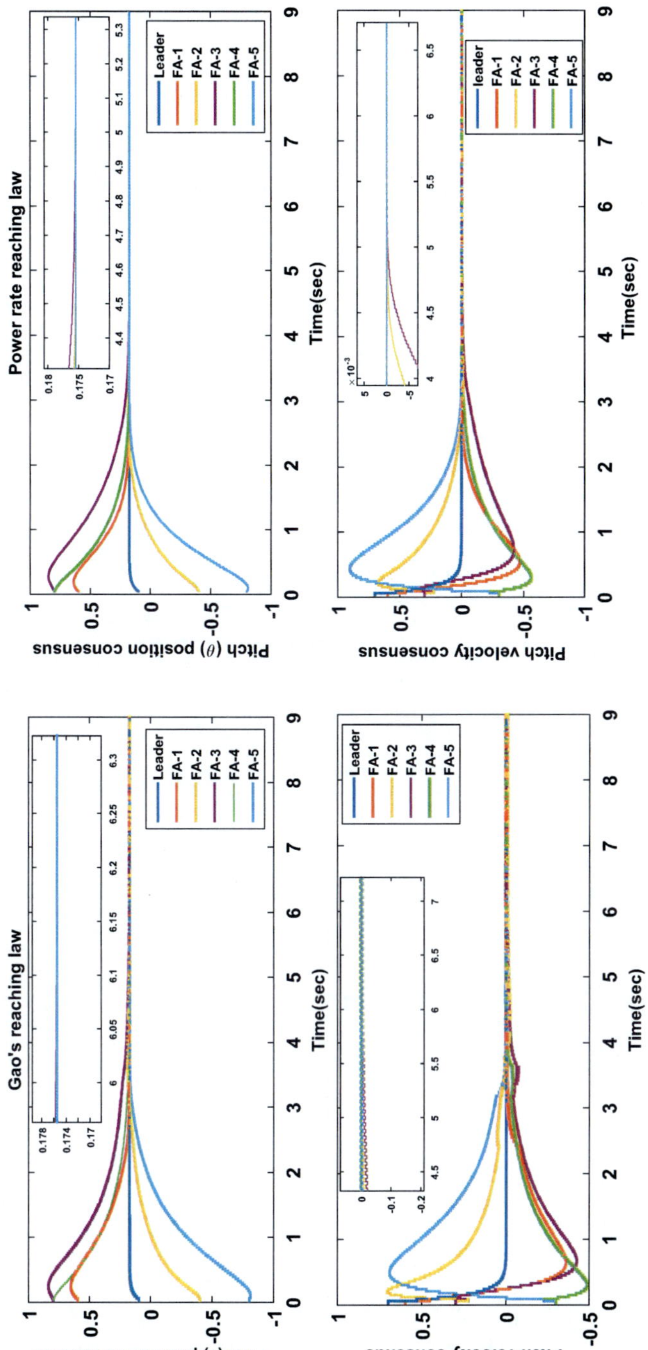

Fig. 4.2 Pitch position and velocity consensus

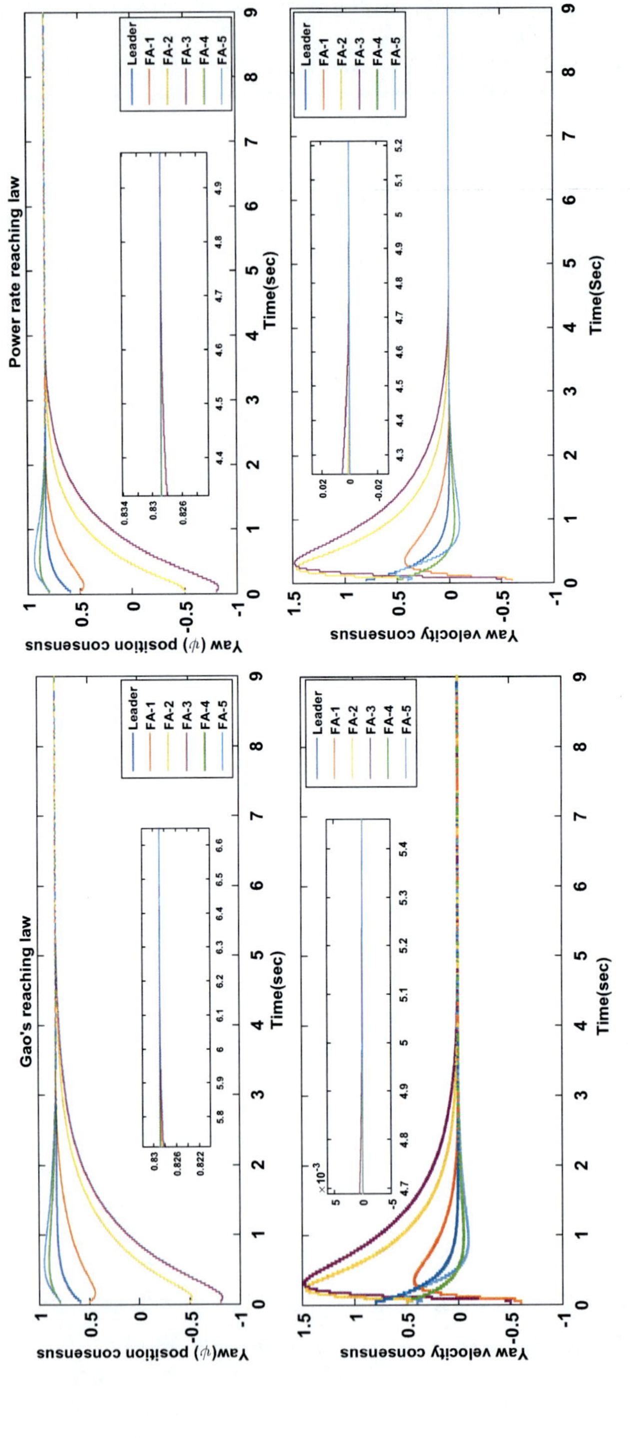

Fig. 4.3 Yaw position and velocity consensus

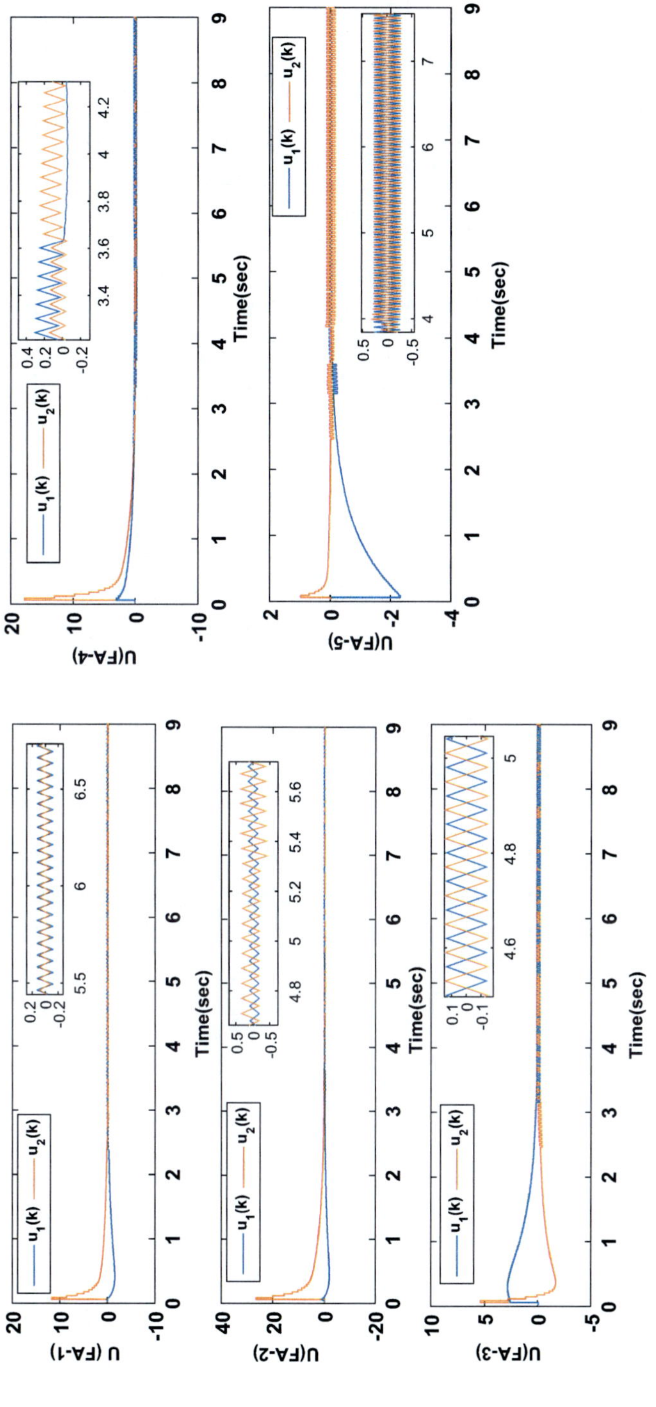

Fig. 4.4 Control effort (u) of each agent using the protocol with Gao's reaching law

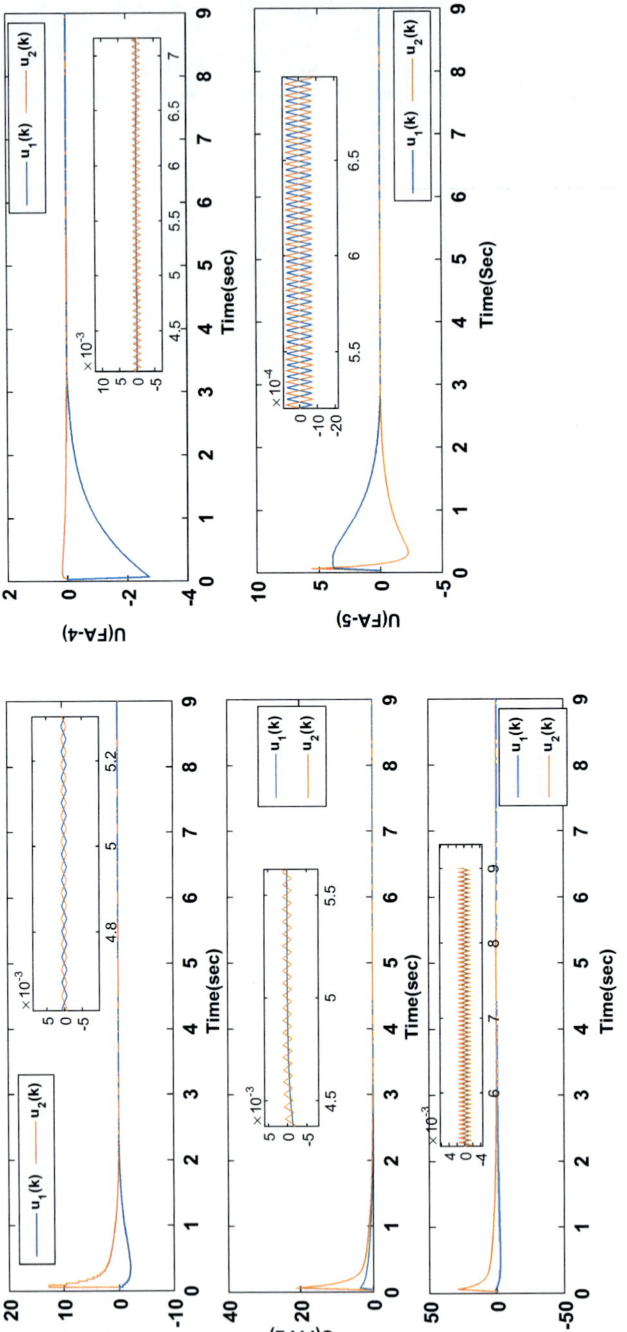

Fig. 4.5 Control effort (u) of each agent using the protocol with Power reaching law

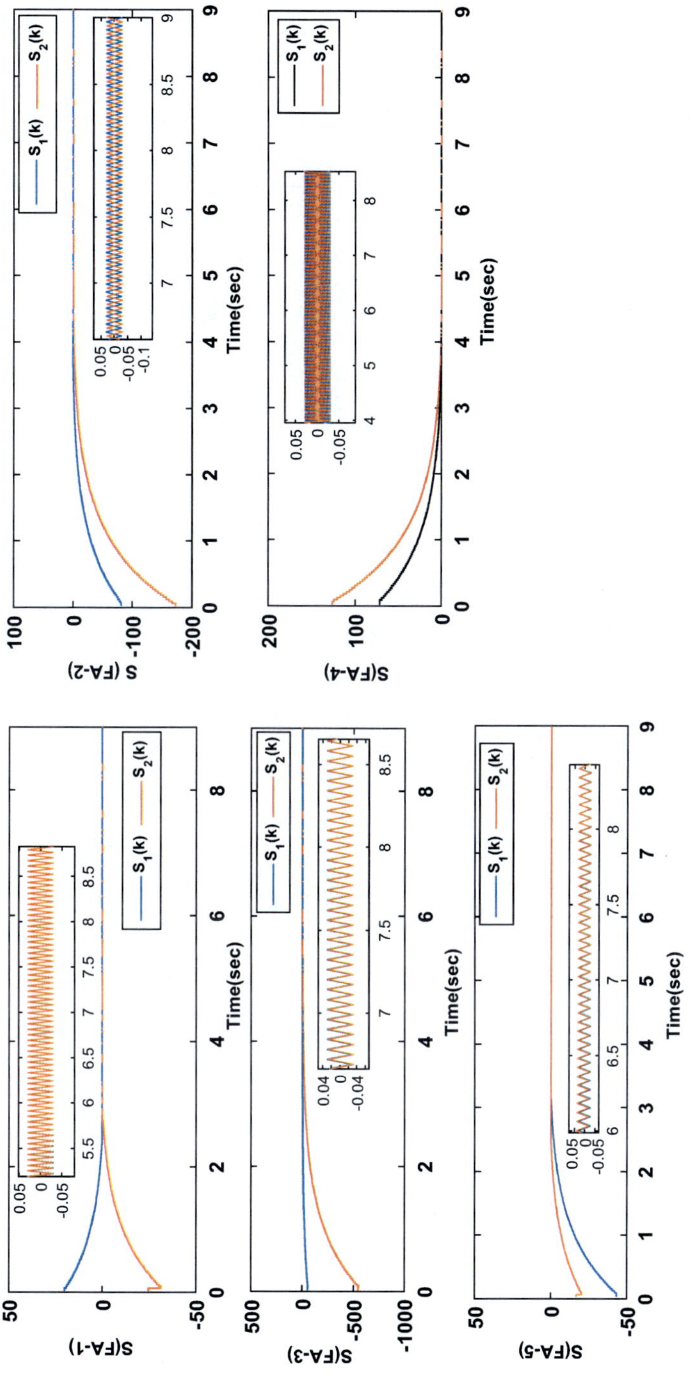

Fig. 4.6 Sliding variable of each agent using Gao's reaching law

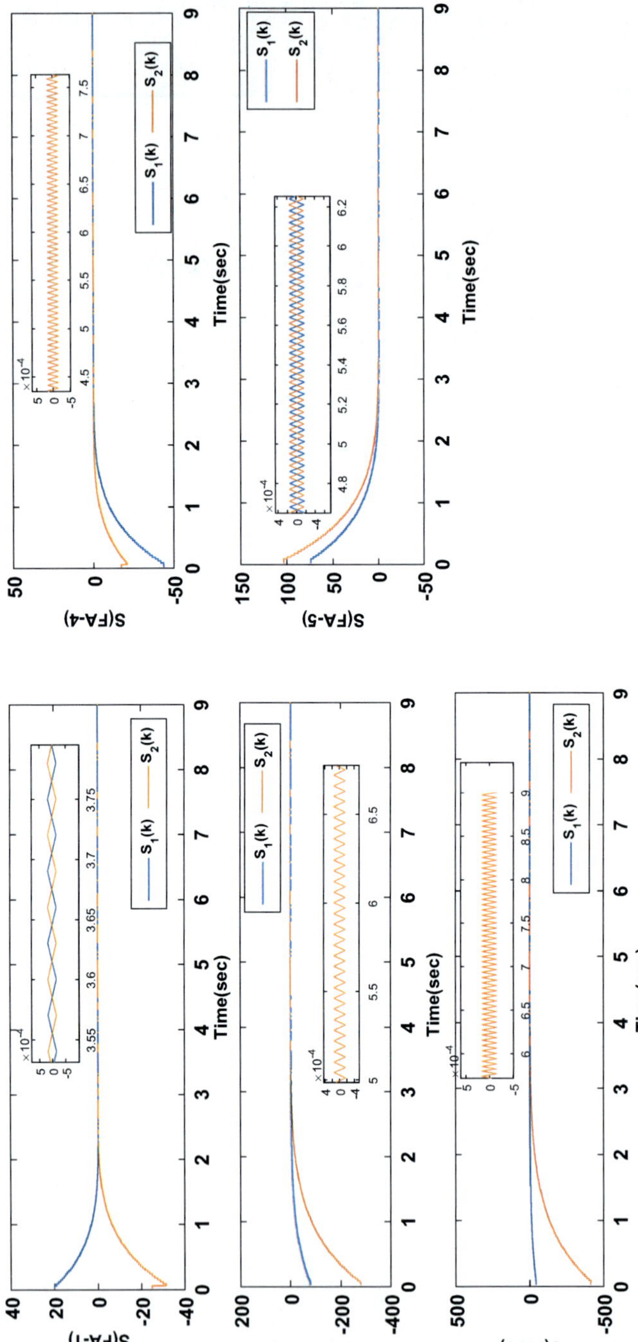

Fig. 4.7 Sliding variable of each agent using Power rate reaching law

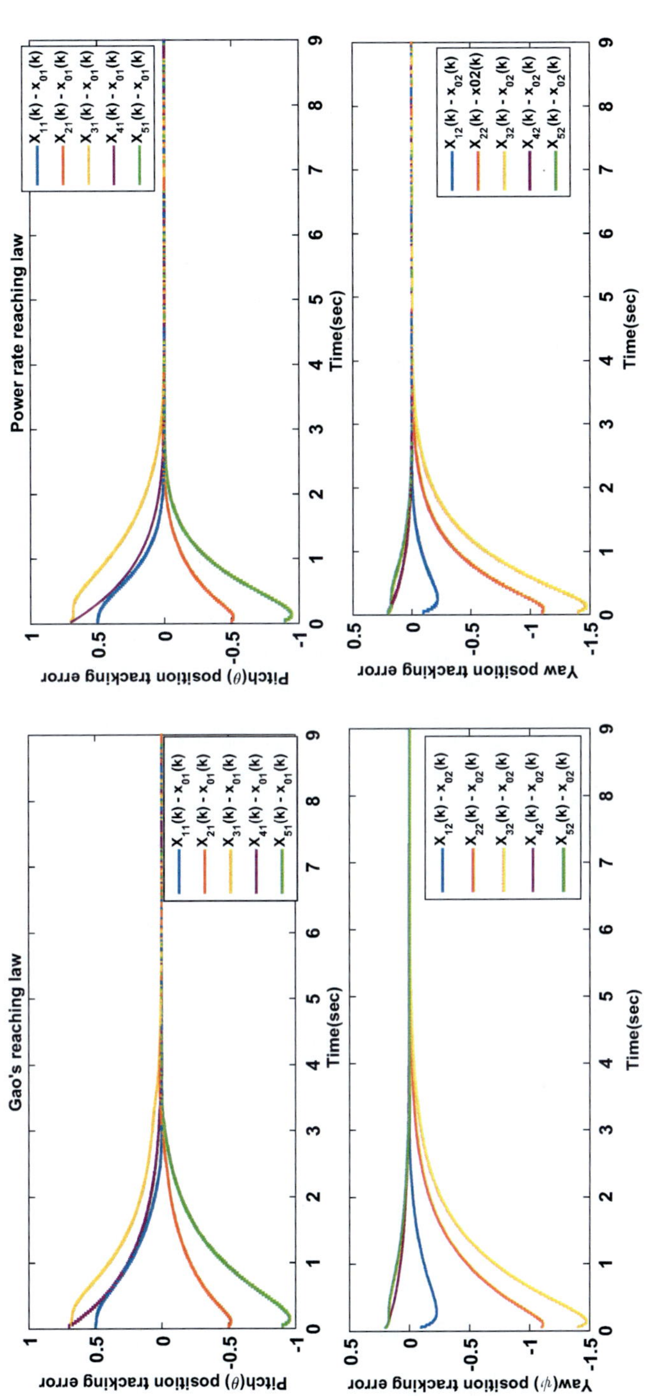

Fig. 4.8 Comparison of pitch and yaw position consensus error

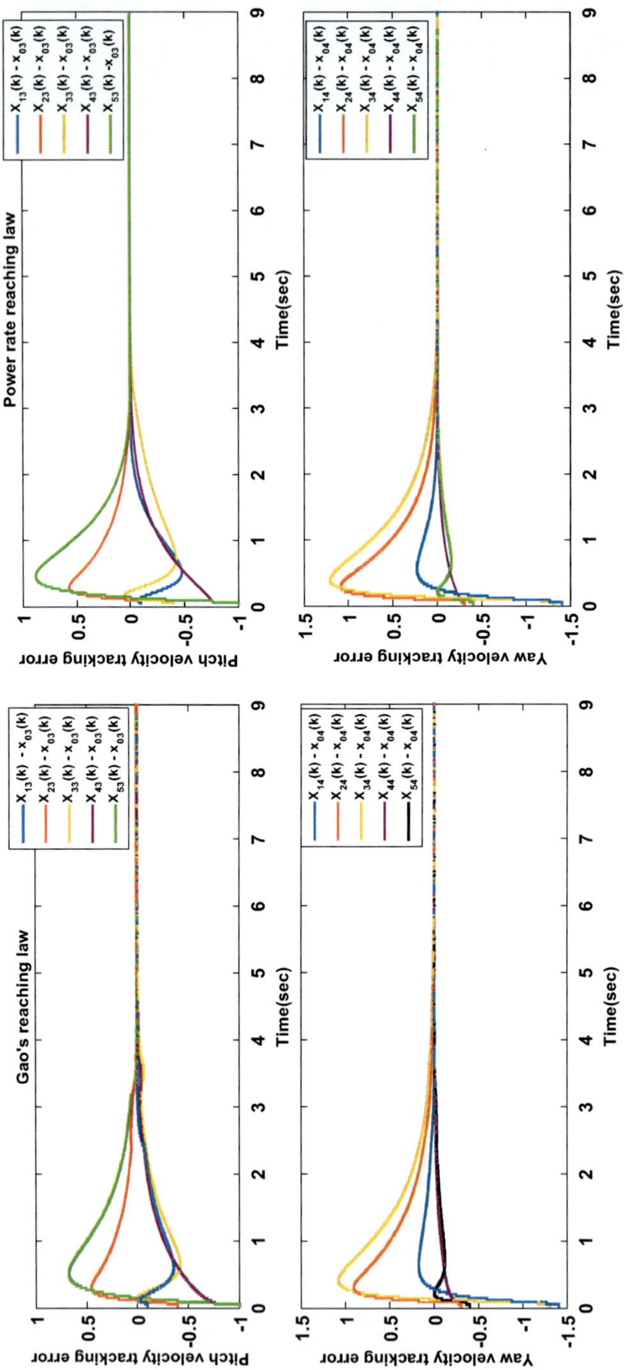

Fig. 4.9 Comparison of Pitch and yaw velocity consensus error

Figure 4.3 shows the comparative study of yaw position and velocity states consensus of the follower with leader states using proposed DSM protocol and found that all the follower agents achieve consensus with the leader in fixed step sizes using switching graph topology.

Figures 4.4 and 4.5 show the consensus effort (u) of the follower agents, which is further applied to individual agent into the overall network of homogeneous DMAS.

Figures 4.6 and 4.7 show the sliding surface (considered as consensus error trajectories) of each agent of DMAS. It crosses first time sliding surface band called as QSMB from any initial condition in fixed time steps and steady within this band and value of this band is varied from –0.03 to +0.03 ($O(\tau)$) calculated using (4.16) in the case of Gao's reaching law and similarly in the case of Power rate reaching law sliding surface remains steadily in $O(\tau^2)$ band once it is on the sliding surface.

Figures 4.8 and 4.9 show the pitch and yaw position and velocity consensus error using two different reaching laws and observed that reached within fixed time steps. It may be noted that the DSM protocol using Power rate reaching law quickly reaches to the zero as compared to DSM protocol using Gao's reaching law.

4.7 Conclusion

In this chapter, two discrete-time sliding mode protocols using Gao's reaching law and Power rate reaching law are proposed for leader-following consensus of discrete multi-agent system configured with a fixed, undirected switching graph topology. Both protocols achieve the leader-following consensus under the influence of switching topology at different intervals. The condition for global stability is also derived using the Lyapunov function. Finally, both protocols are validated in simulation for the 2-DOF helicopter system.

References

1. Patel, K., Mehta, A.: Discrete-time sliding mode protocols for leader-following consensus of discrete multi-agent system with switching graph topology. Eur. J. Control **51**, 65–75 (2020). Jan.
2. Ren, C., Chen, C.L.P.: Sliding mode leader-following consensus controllers for second-order non-linear multi-agent systems. IET Control Theory Appl. **9**(10), 1544–1552 (2015)
3. Ni, W., Cheng, D.: Leader-following consensus of multi-agent systems under fixed and switching topologies. Syst. Control Lett. **59**, 209–217 (2010). Mar.
4. Patel, K., Mehta, A.: Discrete-time sliding mode control for leader following discrete-time multi-agent system. In: IECON 2018 - 44th Annual Conference of the IEEE Industrial Electronics Society, pp. 2288–2292 (2018)
5. Patel, K., Mehta, A.: Discrete higher order sliding mode protocol for leader-following consensus of heterogeneous discrete multi-agent system. In: Lecture Notes in Electrical Engineering, pp. 1–10. Springer, Singapore (2019)

6. Patel, K., Mehta, A.: Discrete-time event-triggered higher order sliding mode control for consensus of 2-dof robotic arms. Eur. J. Control (2020)
7. Varga, R.S.: Geršgorin and His Circles. Springer, Berlin (2004)
8. Gao, W., Wang, Y., Homaifa, A.: Discrete-time variable structure control systems. IEEE Trans. Indust. Electron. **42**, 117–122 (1995)
9. Zhao, Y.-X., Wu, T., Ma, Y.: A double power reaching law of sliding mode control based on neural network. Math. Prob. Eng. **2013**, 1–9 (2013)

Chapter 5
Discrete-Time Higher Order Sliding Mode Protocols for Leader-Following Consensus of Homogeneous Discrete Multi-Agent System

5.1 Introduction of Leader-Following Homogeneous DMAS

Let us define the leader agent dynamical system as

$$x_0(k+1) = \tilde{E}x_0(k) + \tilde{F}u_0(k), \tag{5.1}$$

where $x_0(k) \in \mathbb{R}^n$ is the state vector of the leader agent.

Problem statement: To design and analyze a DSM protocol using reaching law and DSTA for global consensus of homogeneous DMAS such that all the follower agents (3.3) in the graph topology attain consensus with the leader agent (5.1) in finite time by exchanging their information with the neighborhood agent.

5.2 DSM Protocols for the Global Consensus of Homogeneous DMAS

In this section, first, we present the background of the relative degree and then the review of discrete-time higher order sliding mode control proposed in [1, 2].

5.2.1 Relative Degree of Higher Order Discrete Sliding Mode Control Having Relative Degree More Than One

The relative degree of discrete-time systems can be easily understood from Definition 2.1 using continuous-time concept. In discrete-time domain, the derivative operator becomes the difference operator.

K. Patel and A. Mehta, *Discrete-Time Sliding Mode Protocols for Discrete Multi-Agent System*, Studies in Systems, Decision and Control 303,
https://doi.org/10.1007/978-981-15-6311-9_5

5.2.2 *Higher Order Discrete-Time Sliding Mode (DSM) Protocol Using Reaching Law Approach for Consensus of Homogeneous DMAS*

The higher order DSM protocol using the reaching law approach for the leader-following consensus of homogeneous DMAS is presented in the form of Theorem 5.1 as follows.

Theorem 5.1 *For the global homogeneous DMAS (3.3) is said to achieve the consensus in finite time using higher order DSM protocol with reaching law is given as*

$$\begin{aligned} u(k) = [-(\sigma_s^{\mathcal{T}}\bar{\Upsilon}\bar{G}\bar{F})^{-1}[\sigma_s^{\mathcal{T}}\bar{\Upsilon}\bar{E}^2X(k) + \sigma_s^{\mathcal{T}}\bar{\Upsilon}(-1_N \otimes (\tilde{E}x_0(k+1)) - \bar{\lambda}^2\tilde{S}_2(k) \\ +\bar{\lambda}\bar{\vartheta} sgn(\tilde{S}_2(k)) + \bar{\vartheta} sgn(\tilde{S}_2(k+1))]], \end{aligned} \tag{5.2}$$

where $(\sigma_s^{\mathcal{T}}\bar{\Upsilon}\bar{G}\bar{F})^{-1}$ *is non-singular.*

Proof Let us define a sliding surface of ith agent for relative degree-2 using local neighborhood error of leader-following network defined in Eq. (3.6) as

$$\tilde{S}_{i_2}(k) = \sigma_{s_i}^{\mathcal{T}}\bar{\delta}_i(k), \tag{5.3}$$

where $\sigma_{s_i}^{\mathcal{T}}$ is the sliding gain to be designed using either the pole placement or LQ approach. In the leader-follower network, global sliding surface is defined using global consensus error (3.9) as

$$\tilde{S}_2(k) = \sigma_s^{\mathcal{T}}\bar{\Lambda}(k), \tag{5.4}$$

where $\sigma_s^{\mathcal{T}} = I_N \otimes \sigma_{s_i}^{\mathcal{T}}$.

Using Eq. (3.9), we may write Eq. (5.4) as

$$\tilde{S}_2(k) = \sigma_s^{\mathcal{T}}\bar{\Upsilon}(X(k) - 1_N \otimes x_0(k)). \tag{5.5}$$

Further,

$$\tilde{S}_2(k+1) = \sigma_s^{\mathcal{T}}\bar{\Upsilon}(X(k+1) - 1_N \otimes x_0(k+1)). \tag{5.6}$$

Substituting Eqs. (3.3) and (5.1) into Eq. (5.6), we may get

$$\tilde{S}_2(k+1) = \sigma_s^{\mathcal{T}}\bar{\Upsilon}(\bar{E}X(k) + \bar{F}(u(k) + \bar{\mathcal{D}}(k)) - 1_N \otimes \tilde{E}_0x_0(k)). \tag{5.7}$$

Now selecting $\sigma_s^{\mathcal{T}}$ such that $\sigma_s^{\mathcal{T}}\bar{\Upsilon}\bar{F} = 0$, so Eq. (5.7) rewritten as

$$\tilde{S}_2(k+1) = \sigma_s^{\mathcal{T}}\bar{\Upsilon}(\bar{E}X(k) - 1_N \otimes \tilde{E}x_0(k)). \tag{5.8}$$

Further, advancing Eq. (5.8)

$$\tilde{S}_2(k+2) = \sigma_s^{\mathcal{T}} \bar{\Upsilon}(\bar{E}^2 X(k) + \bar{E}\bar{F}(u(k) + \bar{\mathcal{D}}(k)) - 1_N \otimes \tilde{E}x_0(k+1)). \quad (5.9)$$

Now inspired from the reaching law proposed in [3] for linear system, let us define consensus reaching law for ith agent as

$$\tilde{S}_i(k+1) = \lambda_i \tilde{S}_i(k) - \vartheta_i sgn(\tilde{S}_i(k)) + \tilde{\mathcal{D}}_{1_i}(k). \quad (5.10)$$

From Eq. (5.10), consensus reaching law for relative degree-2 is defined as

$$\tilde{S}_{i_2}(k+2) = \lambda_i^2 \tilde{S}_{i_2}(k) - \lambda_i \vartheta_i sgn(\tilde{S}_{i_2}(k)) - \vartheta_i sgn(\tilde{S}_{i_2}(k+1)) + \tilde{\mathcal{D}}_{2_i}(k) (5.11)$$

From Eq. (5.11), we may write the global consensus reaching law for global homogeneous DMAS as

$$\tilde{S}_2(k+2) = \bar{\lambda}^2 \tilde{S}_2(k) - \bar{\lambda}\bar{\vartheta} sgn(\tilde{S}_2(k)) - \bar{\vartheta} sgn(\tilde{S}_2(k+1)) + \tilde{\mathcal{D}}_2(k), \quad (5.12)$$

where $\bar{\vartheta} = [\vartheta_1, \vartheta_2, \vartheta_3, ...\vartheta_N] \in \mathbb{R}^{nN} > 0$, $\bar{\lambda} = [\lambda_1, \lambda_2, \lambda_3, ...\lambda_N] \in \mathbb{R}^{nN} > 0$, $0 < (1 - \bar{\lambda}\tau) < 1$, $\tilde{\mathcal{D}}_2(k) = \sigma_s^{\mathcal{T}} \bar{\Upsilon} \bar{E} \bar{H} \tilde{\mathcal{D}}(k)$, and $|\tilde{\mathcal{D}}_2(k)| < \tilde{D}_m$.

Comparing Eqs. (5.9) and (5.12), we may write

$$\begin{aligned} &\sigma_s^{\mathcal{T}} \bar{\Upsilon}(\bar{E}^2 X(k) + \bar{E}\bar{F}(u(k) + \bar{\mathcal{D}}(k)) - 1_N \otimes \tilde{E}x_0(k+1)) = \\ &\quad \bar{\lambda}^2 \tilde{S}_2(k) - \bar{\lambda}\bar{\vartheta} sgn(\tilde{S}_2(k)) - \bar{\vartheta} sgn(\tilde{S}_2(k+1)) + \tilde{\mathcal{D}}_2(k). \end{aligned} \quad (5.13)$$

Further Eq. (5.13) can be expressed in terms of the protocol for the global consensus of DMAS as

$$\begin{aligned} u(k) = &[-(\sigma_s^{\mathcal{T}} \bar{\Upsilon} \bar{G} \bar{F})^{-1} [\sigma_s^{\mathcal{T}} \bar{\Upsilon} \bar{E}^2 X(k) + \sigma_s^{\mathcal{T}} \bar{\Upsilon}(-1_N \otimes (\tilde{E}x_0(k+1)) - \bar{\lambda}^2 \tilde{S}_2(k) \\ &+ \bar{\lambda}\bar{\vartheta} sgn(\tilde{S}_2(k)) + \bar{\vartheta} sgn(\tilde{S}_2(k+1))]]. \end{aligned} \quad (5.14)$$

This completes the proof.

To obtain the optimal performance of sliding mode protocol, let's consider cost function for finding the sliding mode gain as

$$\bar{J} = \sum_{k=0}^{\infty} (\bar{\Lambda}(k)^{\mathcal{T}} \bar{Q} \bar{\Lambda}(k) + u(k)^{\mathcal{T}} \bar{R} u(k)), \quad (5.15)$$

where, $\bar{Q} = I_N \otimes Q$ and $\bar{R} = I_N \otimes R$ are positive semi-definite and positive definite function respectively. $\bar{Q}$ and $\bar{R}$ represent the relative significance of the global error variation and control energy consumption, respectively. The optimal gain vector $\sigma_s^{\mathcal{T}}$ which minimizes the cost function $\bar{J}$ in Eqn. (5.15) is given by

$$\sigma_s^{\mathcal{T}} = \bar{R}^{-1} \bar{F} \bar{P}. \quad (5.16)$$

The symmetric matrix $\bar{P}$ is the solution of the Riccati equation as

$$\bar{E}^{\mathcal{T}}\bar{P} + \bar{P}\bar{E} - \bar{P}\bar{F}\bar{R}^{-1}\bar{F}^{\mathcal{T}}\bar{P} + \bar{Q} = 0, \tag{5.17}$$

where, $\bar{P} = I_N \otimes P$.

Further, the closed loop dynamic using protocol (5.14) is given as

$$X_2(k+1) = \bar{E}(I_{nN} - \bar{F}(\sigma_s^{\mathcal{T}}\bar{\Upsilon}\bar{E}\bar{F})^{-1}\sigma_s^{\mathcal{T}}\bar{E})X(k). \tag{5.18}$$

The ultimate band Δ_{2_i} of the sliding surface $\tilde{S}_{i_2}(k)$ for ith agent is obtained using Eq. (5.11) by considering $\tilde{S}_{i_2}(k+2) = \Delta_{2_i}(k)$ and maximum disturbance $\tilde{\mathcal{D}}_{2_i}(k) = D_{m_i}(k)$

$$\begin{aligned}\tilde{S}_{i_2}(k+2) = \lambda_i^2\tilde{S}_{i_2}(k) - \lambda_i\vartheta_i sgn(\tilde{S}_{i_2}(k)) - \vartheta_i sgn(\tilde{S}_{i_2}(k+1)) \\ +\tilde{\mathcal{D}}_{2_i}(k).\end{aligned} \tag{5.19}$$

Now using Eq. (5.19), the ultimate band in which consensus error trajectory remains steadily can be derived by considering $\Delta_{2_i} = \tilde{S}_{i_2}(k)$ as

$$\Delta_{2_i} = \frac{(1-\lambda_i)\vartheta_i + D_{m_i}(k)}{1-\lambda_i^2}. \tag{5.20}$$

5.2.3 *Higher Order DSM Protocol Using DSTA for the Global Consensus of Homogeneous DMAS*

In this section, we propose higher order DSM protocol using DSTA for the leader-following consensus of homogeneous DMAS in the form of Theorem 5.2 as given below.

Theorem 5.2 *For the global homogeneous DMAS (3.3) is said to achieve the consensus in finite time using DHSM protocol with DSTA is given as*

$$\begin{aligned}u(k) = [-(\sigma_s^{\mathcal{T}}\bar{\Upsilon}\bar{F})^{-1}[\sigma_s^{\mathcal{T}}\bar{\Upsilon}\bar{E}X(k) + \sigma_s^{\mathcal{T}}\bar{\Upsilon}(-1_N \otimes x_0(k+1))]] \\ -\bar{\mathcal{D}}(k) - h\sqrt{|\tilde{S}(k)|}sgn(\tilde{S}(k)) + w(k),\end{aligned} \tag{5.21}$$

where $(\sigma_s^{\mathcal{T}}\bar{\Upsilon}\bar{F})^{-1}$ is non-singular.

Proof Let us define a sliding surface of ith agent using local neighborhood error of leader-following network defined in Eq. (3.6) as

$$\tilde{S}_i(k) = \sigma_{s_i}^{\mathcal{T}}\bar{\delta}_i(k), \tag{5.22}$$

where $\sigma_{s_i}^{\mathcal{T}}$ is the sliding mode gain to be identified using LQ technique. In the leader-follower network for global system (3.3), the global sliding surface can be written as

$$\tilde{S}(k) = \sigma_s^{\mathcal{T}} \bar{\Lambda}(k). \tag{5.23}$$

Using Eq. (3.9), we may write

$$\tilde{S}(k) = \sigma_s^{\mathcal{T}} \bar{\Upsilon}(X(k) - 1_N \otimes x_0(k)). \tag{5.24}$$

Further,

$$\tilde{S}(k+1) = \sigma_s^{\mathcal{T}} \bar{\Upsilon}(X(k+1) - 1_N \otimes x_0(k+1)). \tag{5.25}$$

Substituting Eqs. (3.3) and (5.1) into Eq. (5.25),

$$\tilde{S}(k+1) = \sigma_s^{\mathcal{T}} \bar{\Upsilon}(\bar{E}X(k) + \bar{F}(u(k) + \bar{\mathcal{D}}(k)) - 1_N \otimes x_0(k+1)). \tag{5.26}$$

The proposed protocol $u(k)$ defined in Eq. (5.21) comprises two parts $u_{eq}(k)$ and $u_s(k)$, where $u_{eq}(k)$ is equivalent control and $u_s(k)$ is the super-twisting control. To derive the $u_{eq}(k)$, let us consider $\tilde{s}(k+1) = 0$ in Eq. (5.26), we may write

$$0 = \sigma_s^{\mathcal{T}} \bar{\Upsilon}(\bar{E}X(k) + \bar{F}(u_{eq}(k) + \bar{\mathcal{D}}(k)) - 1_N \otimes x_0(k+1)). \tag{5.27}$$

Further,

$$u_{eq}(k) = [-(\sigma_s^{\mathcal{T}} \bar{\Upsilon} \bar{F})^{-1}[\sigma_s^{\mathcal{T}} \bar{\Upsilon} \bar{E} X(k) + \sigma_s^{\mathcal{T}} \bar{\Upsilon}(-1_N \otimes x_0(k+1))]] - \bar{\mathcal{D}}(k). \tag{5.28}$$

Now [2], let us define discrete super-twisting algorithm based discrete sliding mode control for the follower agent as

$$\begin{aligned} u_s(k) &= -h\sqrt{\mid \tilde{S}(k) \mid} sgn(\tilde{S}(k)) + w(k) \\ w(k+1) &= w(k) - \mathcal{F} \tau sgn(\tilde{S}(k)), \end{aligned} \tag{5.29}$$

where $h = [h_1, h_2, h_3, ...h_N] \in \mathbb{R}^{nN}$, $\mathcal{F} = [\mathcal{F}_1, \mathcal{F}_2, \mathcal{F}_3, ...\mathcal{F}_N] \in \mathbb{R}^{nN}$ are gain parameters, and τ is the sampling time.

Using Eqs. (5.28) and (5.29), we may get

$$\begin{aligned} u(k) = {} & [-(\sigma_s^{\mathcal{T}} \bar{\Upsilon} \bar{F})^{-1}[\sigma_s^{\mathcal{T}} \bar{\Upsilon} \bar{E} X(k) + \sigma_s^{\mathcal{T}} \bar{\Upsilon}(-1_N \otimes x_0(k+1))]] \\ & -\bar{\mathcal{D}}(k) - h_1\sqrt{\mid \tilde{S}(k) \mid} sgn(\tilde{S}(k)) + w(k). \end{aligned} \tag{5.30}$$

This completes the proof.

Next, the condition for global stability of the homogeneous DMAS with the proposed DSM protocol is derived in the form of Theorem 5.3 as follows.

Theorem 5.3 *The Lyapunov-based global stability of leader-following consensus of homogeneous DMAS using protocol defined in (5.30) is guaranteed if the global error dynamics defined in (3.9) drives toward the global sliding surface (5.24) and maintain on it for gain* $h, \mathcal{F}\tau > 0$*, provided the following conditions hold true:*

$$0 \leq \bar{\alpha} < \tilde{S}^{\mathcal{T}}(k)\tilde{S}(k), \tag{5.31}$$

where $\bar{\alpha} = [-h\sqrt{\mid \tilde{S}(k) \mid} sgn(\tilde{S}(k)) + w(k)]^{\mathcal{T}} * [-h\sqrt{\mid \tilde{S}(k) \mid} sgn(\tilde{S}(k)) + w(k)]$.

Proof Let us define global sliding surface using Eq. (5.24)

$$\tilde{S}(k) = \sigma_s^{\mathcal{T}} \bar{\Upsilon}(X(k) - 1_N \otimes x_0(k)). \tag{5.32}$$

Further,

$$\tilde{S}(k+1) = \sigma_s^{\mathcal{T}} \bar{\Upsilon}(X(k+1) - 1_N \otimes x_0(k+1))). \tag{5.33}$$

Let us select the Lyapunov function as

$$\bar{\mathcal{V}}(k) = \tilde{S}^{\mathcal{T}}(k)\tilde{S}(k). \tag{5.34}$$

Writing the forward derivative function of Eq. (5.34)

$$\Delta\mathcal{V}_s(k) = \mathcal{V}(k+1) - \mathcal{V}(k), \tag{5.35}$$

$$\Delta\mathcal{V}_s(k) = \tilde{S}^{\mathcal{T}}(k+1)\tilde{S}(k+1) - \tilde{S}^{\mathcal{T}}(k)\tilde{S}(k). \tag{5.36}$$

For stability, it is required to ensure $\Delta\mathcal{V}_s(k) < 0$. Now substituting $\tilde{S}(k+1)$ in Eq. (5.33) into Eq. (5.36), we get

$$\begin{aligned}\Delta\mathcal{V}_s(k) = {} & [\sigma_s^{\mathcal{T}} \bar{\Upsilon}(X(k+1) - 1_N \otimes x_0(k+1))]^{\mathcal{T}} [\sigma_s^{\mathcal{T}} \bar{\Upsilon}(X(k+1) - \\ & 1_N \otimes x_0(k+1))] - \tilde{S}^{\mathcal{T}}(k)\tilde{S}(k).\end{aligned} \tag{5.37}$$

Substituting $X(k+1)$ and $x(k+1)$ from Eqs. (3.3) and (5.1),

$$\begin{aligned}\Delta\mathcal{V}_s(k) = {} & [\sigma_s^{\mathcal{T}} \bar{\Upsilon}[(\bar{E}X(k) + \bar{F}(u(k) + \bar{\mathcal{D}}(k)) - 1_N \otimes x_0(k+1)]]^{\mathcal{T}} [\sigma_s^{\mathcal{T}} \bar{\Upsilon}[(\bar{E}X(k) \\ & + \bar{F}(u(k) + \bar{\mathcal{D}}(k)) - 1_N \otimes x_0(k+1))]] - \tilde{S}^{\mathcal{T}}(k)\tilde{S}(k).\end{aligned} \tag{5.38}$$

Substituting the protocol defined in Eq. (5.30) into Eq. (5.38), we may get

$$\Delta\mathcal{V}_s(k) = [-h\sqrt{\mid \tilde{S}(k) \mid} sgn(\tilde{S}(k)) + w(k)]^{\mathcal{T}} * [-h\sqrt{\mid \tilde{S}(k) \mid} sgn(\tilde{S}(k)) + w(k)] - \tilde{S}^{\mathcal{T}}(k)\tilde{S}(k), \tag{5.39}$$

and denoting,

$$\bar{\alpha} = [-h\sqrt{|\ \tilde{S}(k)\ |}sgn(\tilde{S}(k)) + w(k)]^{\mathcal{T}} * [-h\sqrt{|\ \tilde{S}(k)\ |}sgn(\tilde{S}(k)) + w(k)].$$

$$\Delta\mathcal{V}_s(k) = \bar{\alpha} - \tilde{S}^{\mathcal{T}}(k)\tilde{S}(k). \tag{5.40}$$

From Eq. (5.40), the term α can be tuned close to zero by selecting the proper value of h and $\mathcal{F}$ such that $\Delta\mathcal{V}_s(k) < 0$. Hence, the global stability of homogeneous DMAS is guaranteed.

This completes the proof.

5.3 Design, Implementation, and Comparison of Higher Order DSM Protocols on 2-DOF Serial Flexible Joint Robotic Arm

5.3.1 System Description and Mathematical Modeling

The 2-Degree-of-Freedom (DOF) Serial Flexible Joint (2DOFSFJ) mechanism is portrayed in Fig. 5.1. The system consists of 2 DC motors driving harmonic gearboxes (zero backlashes) and a two-bar serial linkage. Both links are rigid. The first link is coupled to the primary drive by means of a versatile joint. It carries at its end the second harmonic drive that is coupled to the second rigid link via another flexible joint. Each motor and each versatile joint are instrumented with optical encoders. Each flexible joint uses two springs which can be modified. A thumbscrew mechanism is accessible to maneuver every spring end to completely different anchor points on its support bars as desired. The system parameters of each joint are given in Table 5.1.

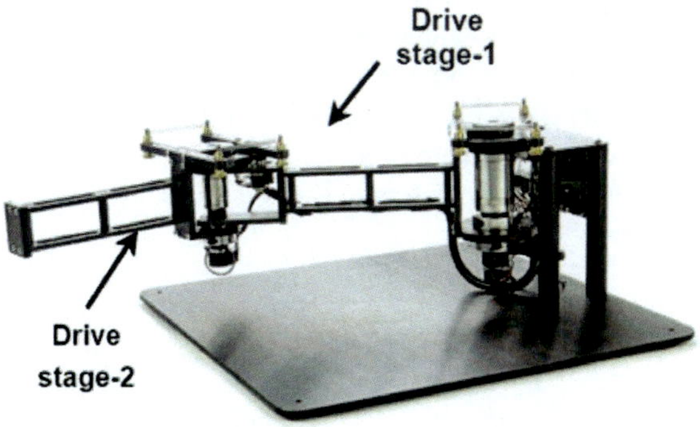

Fig. 5.1 2-DOF Serial Flexible Joint Robotic arm

Table 5.1 System parameters of 2DOFSFJ robotic arm

System parameters	
First drive stage	
I_m	Motor armature current (A)
K_t	Torque constant (N.m/A)
τ	Torque produced at the load shaft (N.m)
θ_{11}	Angular position (rad)
$\dot{\theta}_{11}$	Angular velocity (rad/s)
θ_{12}	1st rigid link absolute angular position link to driving stage-2
$\dot{\theta}_{12}$	1st rigid link absolute angular velocity link to driving stage-2
J_{11}	Moment of inertia (kg.m^2)
J_{12}	Moment of inertia (compounded with stage-2) (kg.m^2)
B_{11}	Viscous damping coefficient (N.m.s/rad)
B_{12}	Viscous damping coefficient (compounded with stage-2) (N.m.s/rad)
K_s	Torsional stiffness constant (N.m/rad)
Second drive stage	
θ_{21}	2nd rigid link absolute angular position link to driving stage-1 (rad)
$\dot{\theta}_{21}$	2nd rigid link absolute angular velocity link to driving stage-1 (rad/s)
θ_{22}	Angular position (rad)
$\dot{\theta}_{22}$	Angular velocity (rad/s)
J_{21}	Moment of inertia(compounded with stage-1) (kg.m^2)
J_{22}	Moment of inertia (kg.m^2)
B_{21}	Viscous damping coefficient (compounded with stage-1) (N.m.s/rad)
B_{22}	Viscous damping coefficient (N.m.s/rad)

Considering 2-DOF serial flexible joint robotic arm to be rigid is modeled by neglecting the motion of actuators and various fictional forces [4]. The dynamical equations of the flexible joints are solved using the Euler–Lagrange method. The entire robotic arm system is divided into two stages: (i) drive stage-1 and (ii) drive stage-2. θ_{12}, θ_{21} represent the angular position of stage-1 with respect to drive stage-2 and stage-2 angular position with respect to stage-1. Similarly, $\dot{\theta}_{12}, \dot{\theta}_{21}$ represent the relative angular velocity of two drive stages [5, 6].

Using linear the Euler–Lagrange method, first, let us write the dynamic equation for driving stage-1 of 2DOFSFL robotic arm system as

$$\ddot{\theta}_{11}(t) = \frac{-K_{s_1}\theta_{11}(t)}{J_{11}} + \frac{K_{s_1}\theta_{12}(t)}{J_{11}} - \frac{B_{11}\dot{\theta}_{11}(t)}{J_{11}} + \frac{k_{t_1}^2 I_{m_1}}{J_{11}}$$

$$\ddot{\theta}_{12}(t) = \frac{-K_{s_1}\theta_{11}(t)}{J_{12}} - \frac{K_{s_1}\theta_{12}(t)}{J_{12}} - \frac{B_{12}\dot{\theta}_{12}(t)}{J_{12}}. \tag{5.41}$$

From Eq. (5.41), the state space model for the driving stage-1 is obtained as

$$\tilde{E}_1 = \begin{bmatrix} 0 & 0 & 1 & 0 \\ 0 & 0 & 0 & 1 \\ \frac{-K_{s_1}}{J_{11}} & \frac{K_{s_1}}{J_{11}} & \frac{-B_{11}}{J_{11}} & 0 \\ \frac{K_{s_1}}{J_{12}} & \frac{-K_{s_1}}{J_{12}} & 0 & -\frac{-B_{12}}{J_{12}} \end{bmatrix} \in \mathbb{R}^{n\times n},$$
$$\tilde{F}_1 = \begin{bmatrix} 0 \; 0 \; \frac{K_{t_1}}{J_{11}} \; 0 \end{bmatrix}^{\mathcal{T}} \in \mathbb{R}^{n\times m}, \quad \tilde{G}_1 = \begin{bmatrix} 1 \; 0 \; 0 \; 0 \end{bmatrix}, \tag{5.42}$$

where $\tilde{E}_1$, $\tilde{F}_1$, and $\tilde{G}_1$ are the system matrix, input matrix, and output matrix, respectively. $\mathcal{T}$ represents the transpose of the matrix. Similarly for the driving stage-2, the dynamics can be written using Euler–Lagrange equations of motion as

$$\ddot{\theta}_{21}(t) = \frac{-K_{s_2}\theta_{21}(t)}{J_{21}} + \frac{K_{s_2}\theta_{22}(t)}{J_{21}} - \frac{B_{21}\dot{\theta}_{21}(t)}{J_{21}} + \frac{k_{t_2}I_{m_2}}{J_{21}}, \tag{5.43}$$

$$\ddot{\theta}_{22}(t) = \frac{K_{s_2}\theta_{21}(t)}{J_{22}} - \frac{K_{s_2}\theta_{22}(t)}{J_{22}} - \frac{B_{22}\dot{\theta}_{22}(t)}{J_{22}}. \tag{5.44}$$

And the state space model for the driving stage-2 is obtained as

$$\tilde{E}_2 = \begin{bmatrix} 0 & 0 & 1 & 0 \\ 0 & 0 & 0 & 1 \\ \frac{-K_{s_2}}{J_{21}} & \frac{K_{s_2}}{J_{21}} & \frac{-B_{21}}{J_{21}} & 0 \\ \frac{K_{s_2}}{J_{22}} & \frac{-K_{s_2}}{J_{22}} & 0 & -\frac{-B_{22}}{J_{22}} \end{bmatrix} \in \mathbb{R}^{n\times n},$$
$$\tilde{F}_2 = \begin{bmatrix} 0 \; 0 \; \frac{K_{t_2}}{J_{21}} \; 0 \end{bmatrix}^{\mathcal{T}} \in \mathbb{R}^{n\times m}, \quad \tilde{G}_2 = \begin{bmatrix} 1 \; 0 \; 0 \; 0 \end{bmatrix}. \tag{5.45}$$

To obtain the discrete-time model, the dynamic models (5.42) and (5.45) for stage-1 and stage-2 are discretized at sampling rate $\tau = 0.002$ and are obtained as

$$\tilde{E}d_1 = \begin{bmatrix} 0.9997 & 0.0002696 & 0.001865 & 1.818\times 10^{-7} \\ 7.81\times 10^{-5} & 0.9999 & 5.028\times 10^{-8} & 0.001999 \\ -0.2634 & 0.2634 & 0.868 & 0.0002695 \\ 0.07809 & -0.07809 & 7.455\times 10^{-5} & 0.9993 \end{bmatrix},$$
$$\tilde{F}d_1 = \begin{bmatrix} 0.0002673 \; 3.545\times 10^{-9} \; 0.2612 \; 7.042e\times 10^{-6} \end{bmatrix},$$
$$\tilde{G}d_1 = \begin{bmatrix} 1 \; 0 \; 0 \; 0 \end{bmatrix}. \tag{5.46}$$

Similarly, for stage-2

$$\tilde{E}d_2 = \begin{bmatrix} 0.9979 & 0.002078 & 0.001739 & 1.416 \times 10^{-6} \\ 0.0007443 & 0.9993 & 4.63 \times 10^{-7} & 0.001994 \\ -1.983 & 1.983 & 0.7499 & 0.002075 \\ 0.7433 & -0.7433 & 0.0006783 & 0.994 \end{bmatrix},$$
$$\tilde{F}d_2 = \begin{bmatrix} 0.0004521 \; 5.827 \times 10^{-8} \; 0.4315 \; 0.0001149 \end{bmatrix},$$
$$\tilde{G}d_2 = \begin{bmatrix} 1 \; 0 \; 0 \; 0 \end{bmatrix}. \tag{5.47}$$

5.3.2 *Experimental Implementation of Higher Order DSM Protocols*

The experimental setup shown in Fig. 5.2 comprises four 2-DOF serial flexible joints having the communication topology as shown in Fig. 5.3. Among four robotic arm systems, three systems including leader are virtual and one system is actual one. The robotic arm indexed as 0 acts as leader and 1, 2, and 3 indexed robotic arms act as followers. In this study, position (θ_{11}) of stage-1 and position (θ_{21}) of stage-2 are to be considered as a single agent parameter for consensus. The simulation and experimental study are carried out using MATLAB R15 interfacing with QUARC software [4]. QUARC is the most suitable software to design, develop, and validate applications in the real-time domain on hardware using MATLAB Simulink.

Square signals with amplitude of 30 (deg) with frequency of 0.1 Hz for stage-1 of 2DOFSFJ and an amplitude of 20 (deg) with frequency of 0.1 Hz for stage-2 are given to the leader agent as a reference. The Laplacian matrix $\tilde{L}$, adjacency matrix $\mathcal{A}$, pinning gain matrix $\tilde{\mathcal{B}}$, and degree matrix $\tilde{D}$ are given as

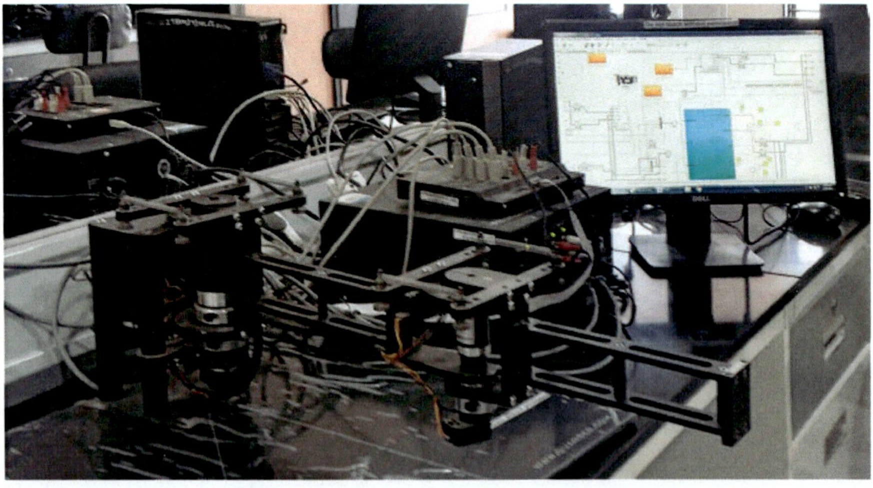

Fig. 5.2 Experimental setup for leader-follower consensus

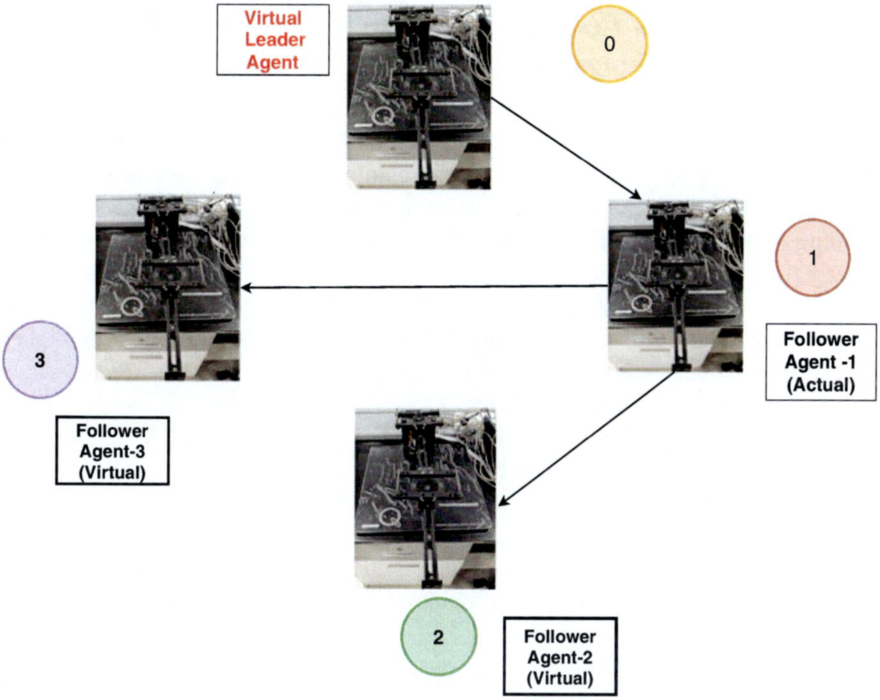

Fig. 5.3 Communication topology

$$\tilde{L} = \begin{bmatrix} 2 & 0 & 0 \\ -1 & 0 & 0 \\ -1 & 0 & 0 \end{bmatrix}, \mathcal{A} = \begin{bmatrix} 0 & 0 & 0 \\ 1 & 0 & 0 \\ 1 & 0 & 0 \end{bmatrix},$$

$$\tilde{\mathcal{B}} = \begin{bmatrix} 1 & 0 & 0 \\ 0 & 0 & 0 \\ 0 & 0 & 0 \end{bmatrix}, \tilde{D} = \begin{bmatrix} 2 & 0 & 0 \\ 0 & 0 & 0 \\ 0 & 0 & 0 \end{bmatrix}. \tag{5.48}$$

The robustness property of the derived protocol for the leader-following consensus of homogeneous DMAS is checked by applying a slowly varying matched disturbance with magnitude $\tilde{\mathcal{D}}_i(k) = 0.0002 * cos(0.01k)$ to ith follower agent.

5.4 Results Discussion

In this section, the simulation and experimental results are discussed using the reaching law approach and DSTA. Further, the results are also compared with Discrete-Time First-Order Sliding Mode (DFSM) protocol for consensus performance.

5.4.1 Simulation Results for Higher Order DSM Protocol with Reaching Law Approach

The protocol defined in Eq. (5.14) is applied to the leader-following homogeneous DMAS comprising 2-DOF flexible joint robotic arm for the consensus. The gain for each ith follower agent λ_i, ϑ_i is chosen as 0.03 and 0.05, respectively. The sliding gain for surface of ith follower agent for the stage-1 and stage-2 is calculated using LQ technique as $\sigma_{s_i 1}^{\mathcal{T}} = [23.8619 \quad -2.1913 \quad 1.7969 \quad 3.2838]$ and $\sigma_{s_i 2}^{\mathcal{T}} = [22.06 \quad -6.62 \ 1.66 \ 0.4056]$, respectively. Figures 5.4 and 5.5 show the simulation and experimental results of position consensus for 2DOFSFJ robotic arm of stage-1 and stage-2, respectively. It is inferred from the results that the follower agents achieve the consensus with the leader.

Figures 5.6 and 5.7 show the simulation and experimental results of the sliding surface of the individual ith agent of homogeneous DMAS for stage-1 of stage-2, respectively. It is inferred from the results that the sliding variable is remaining in the band.

Figures 5.8 and 5.9 show the simulation and experimental results of the protocol(u) of the individual ith agent of homogeneous DMAS for stage-1 and stage-2, respectively.

5.4.2 Simulation Results for Higher Order DSM Protocol Using DSTA

The protocol defined in Eq. (5.30) is applied to the leader-following homogeneous DMAS comprising 2-DOF flexible joint robotic arm for the consensus. The gains for the follower agents are chosen as $h = [0.03 \ 0.03 \ 0.03]$, $\mathcal{F} = [0.015 \ 0.015 \ 0.015]$, respectively. The sliding gains of ith follower agent for the stage-1 and stage-2 are

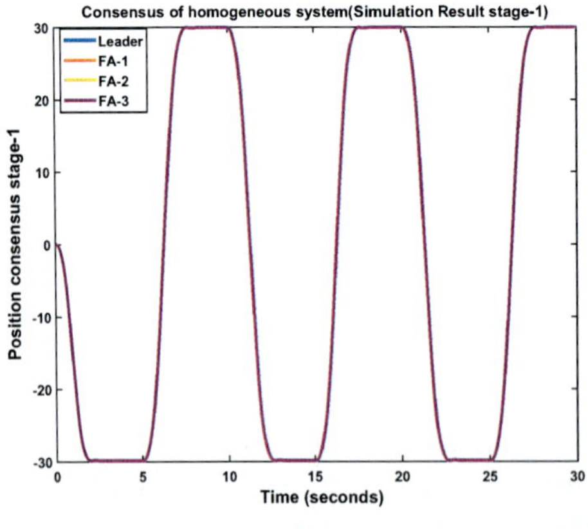

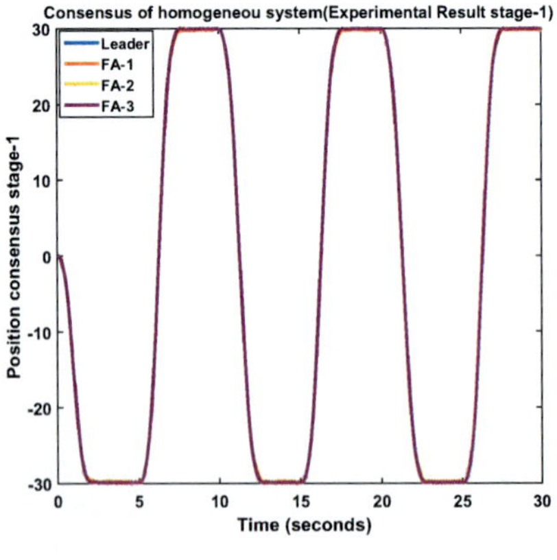

Fig. 5.4 Position consensus (θ_{11}) of 2DOFSFJ robotic arm stage-1

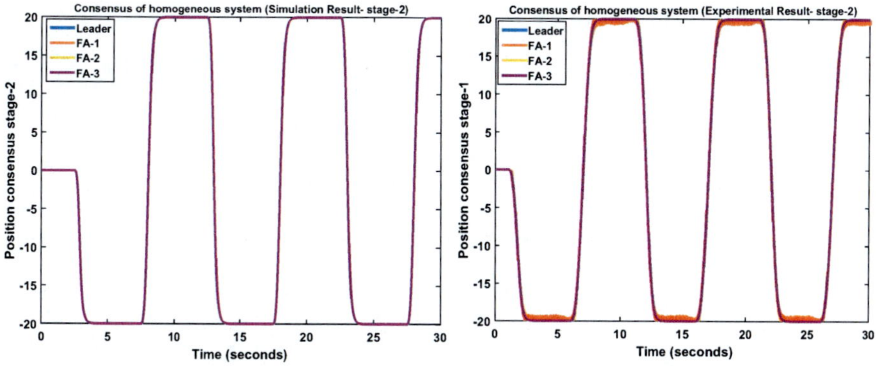

Fig. 5.5 Position consensus (θ_{21}) of 2DOFSFJ robotic arm stage-2

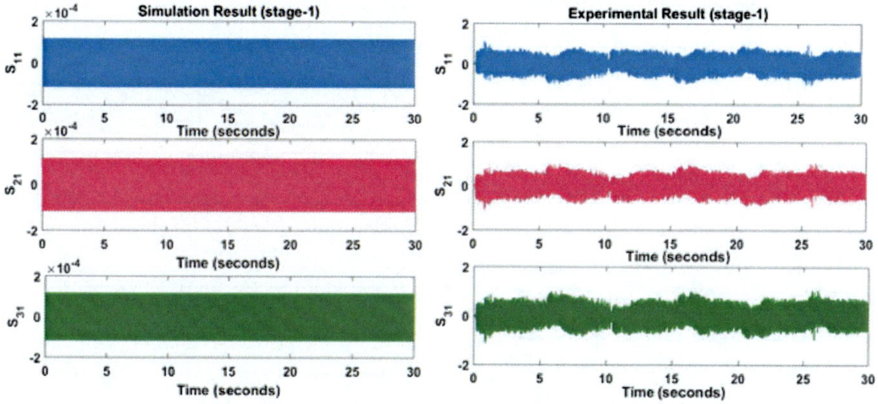

Fig. 5.6 Sliding variable of individual agent of 2DOFSFJ robotic arm stage-1

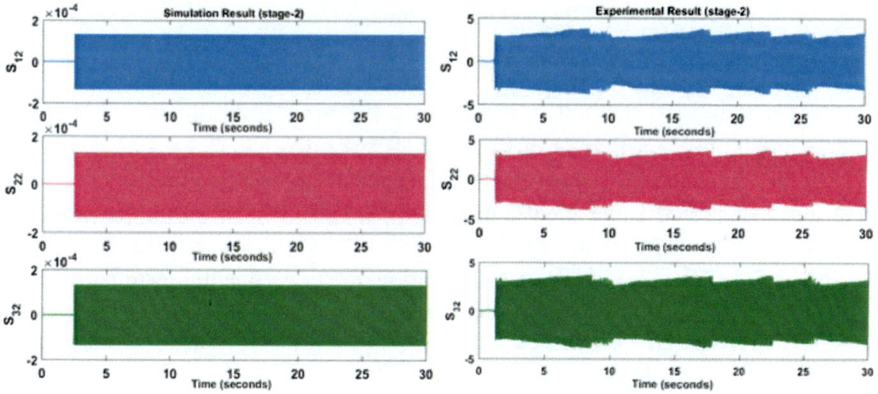

Fig. 5.7 Sliding variable of individual agent of 2DOFSFJ robotic arm stage-2

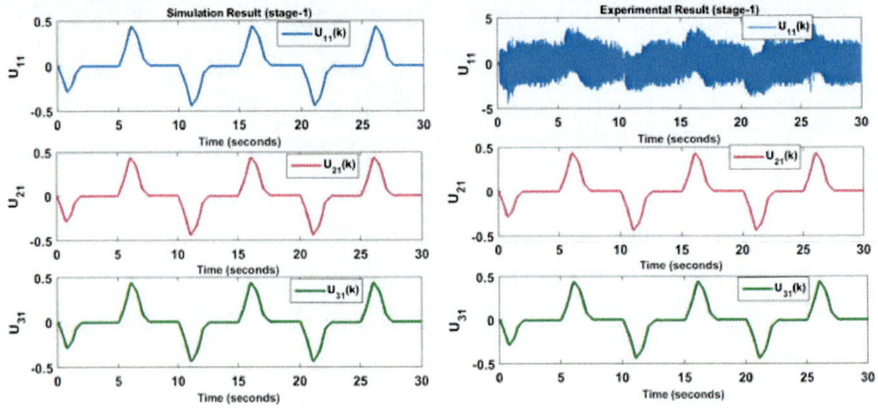

Fig. 5.8 Control effort (u) of 2DOFSFJ robotic arm stage-1

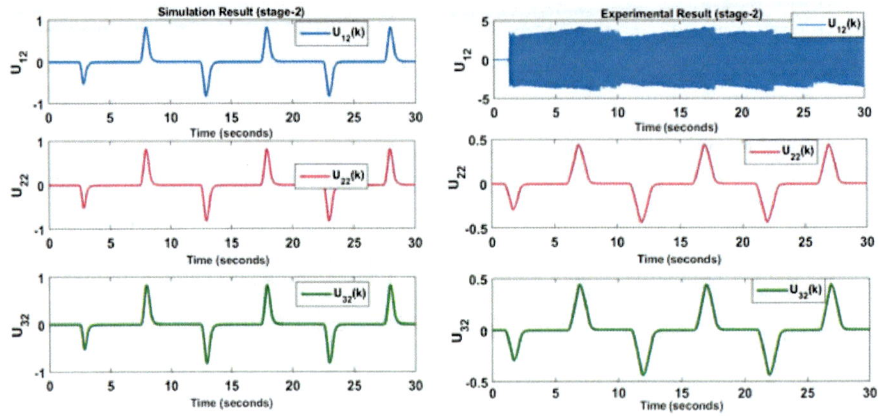

Fig. 5.9 Control effort (u) of 2DOFSFJ robotic arm stage-2

calculated using LQ method as $\sigma^{\mathcal{T}}_{s_i1} = [31.5313 \quad -5.2904 \quad 2.9705 \quad 3.8968]$ and $\sigma^{\mathcal{T}}_{s_i2} = [19.9708 \quad -6.1516 \quad 1.6612 \quad 0.3431]$, respectively.

Figures 5.10 and 5.11 show the simulation and experimental results of position consensus for stage-1 and stage-2 using DSTA. It is observed from the results that all the follower agents attained the consensus with the leader in finite time.

Figures 5.12 and 5.13 show the simulation and experimental results of the sliding surface of ith follower agent for homogeneous DMAS of stage-1 and stage-2, respectively. It is inferred from the results that we are getting less chattering using DSTA compared to DFSMC but higher than the reaching law approach.

Figures 5.14 and 5.15 show the protocol of the individual follower agent for stage-1 and stage-2, respectively, which is further applied to individual agent into the same network.

The proposed protocols are compared with DFSM protocol given in Eq. (2.51) as

$$u(k) = [-(\sigma_s^{\mathcal{T}} \bar{\Upsilon} \bar{F})^{-1}[\sigma_s^{\mathcal{T}} \bar{\Upsilon} \bar{E} X(k) + \sigma_s^{\mathcal{T}} \bar{\Upsilon}(-1_N \otimes \tilde{E}_0 x_0(k)) \\ -(1 - \bar{\lambda}\tau)\tilde{S}(k) + \bar{\vartheta}\tau sgn(\tilde{S}(k))]] - \bar{\mathcal{D}}(k). \quad (5.49)$$

The sliding gains for surface of ith follower agent for the stage-1 and stage-2 are calculated using LQ technique as $\sigma_{s_i1}^{\mathcal{T}} = [29.8664 \ \ -5.9043 \ \ 2.9685 \ \ -0.0043]$ and $\sigma_{s_i2}^{\mathcal{T}} = [21.4614 \ \ -17.7421 \ \ 1.1829 \ \ 0.4598]$, respectively.

Figures 5.16, 5.17, 5.18 and 5.19 show the simulation and experimental results of position consensus and sliding surface for stage-1 and stage-2, respectively. It is inferred from the results that follower agents attained the consensus with the leader agent in finite time step and the chattering is more in DFSM protocol as compared to proposed protocols.

Figures 5.20 and 5.21 show the protocol of follower agent for stage-1 and stage-2, respectively, which further applied to an individual agent in the same network of homogeneous DMAS for the consensus. From results, it is observed that pro-

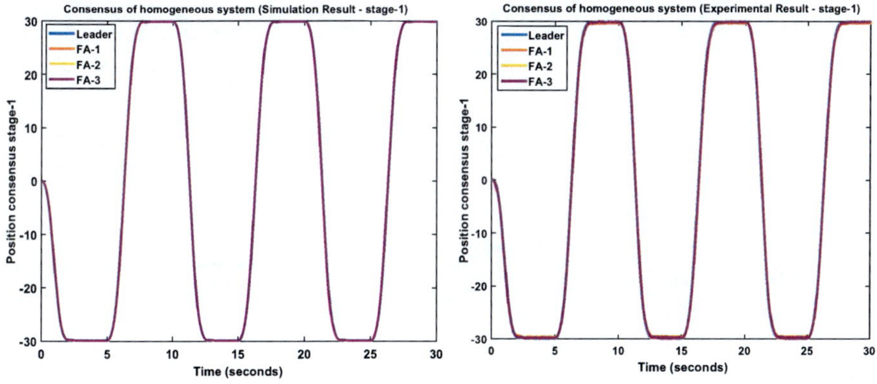

Fig. 5.10 Position consensus (θ_{11}) of 2DOFSFJ robotic arm stage-1

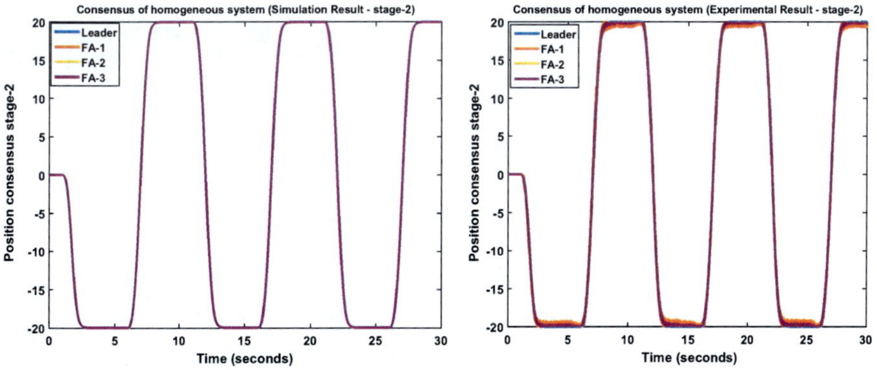

Fig. 5.11 Position consensus (θ_{21}) of 2DOFSFJ robotic arm stage-2

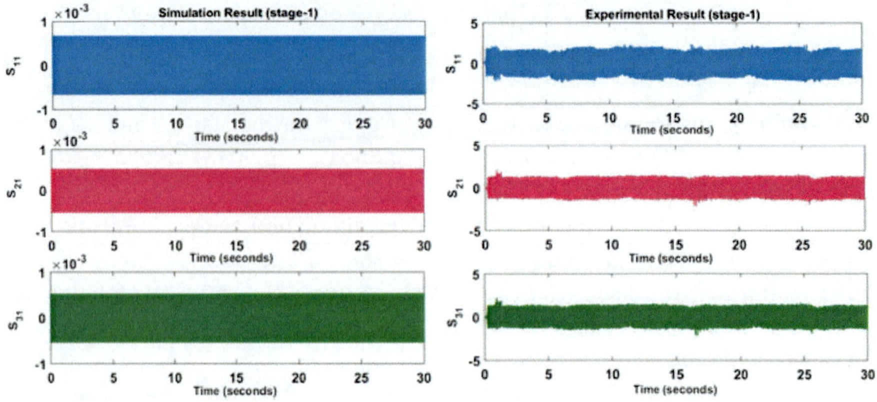

Fig. 5.12 Sliding variable of individual agent of 2DOFSFJ robotic arm stage-1

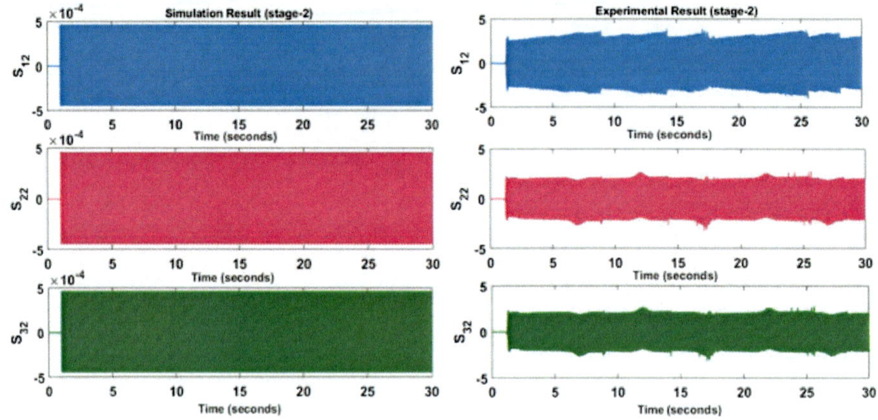

Fig. 5.13 Sliding variable of individual agent of 2DOFSFJ robotic arm stage-2

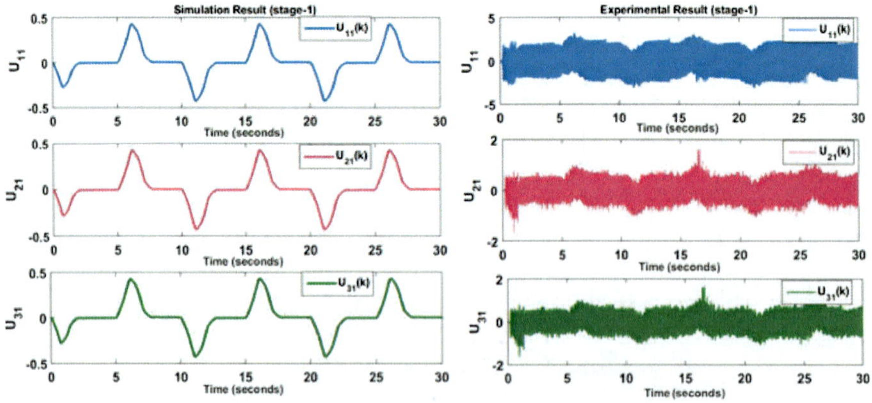

Fig. 5.14 Control effort (u) of 2DOFSFJ robotic arm stage-1

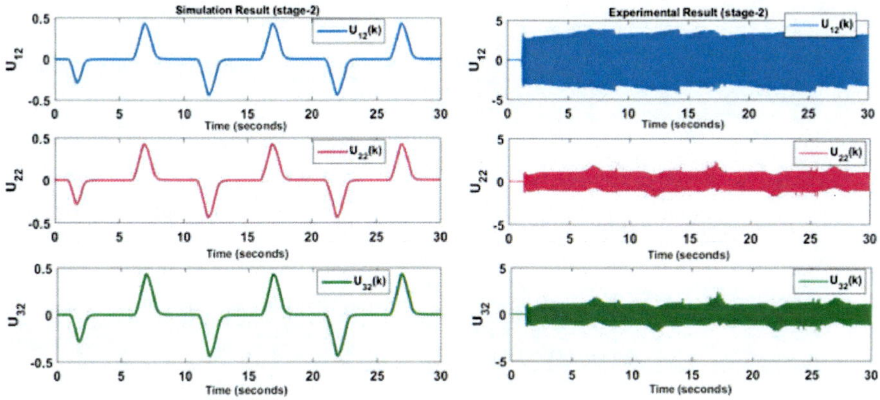

Fig. 5.15 Control effort (u) of 2DOFSFJ robotic arm stage-2

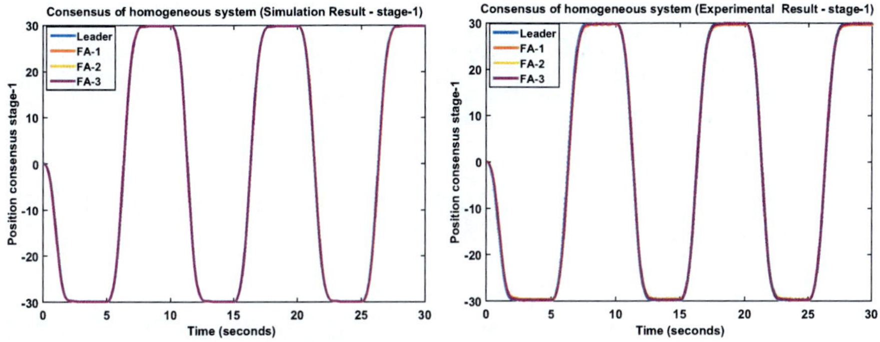

Fig. 5.16 Position consensus (θ_{11}) of 2DOFSFJ robotic arm stage-1

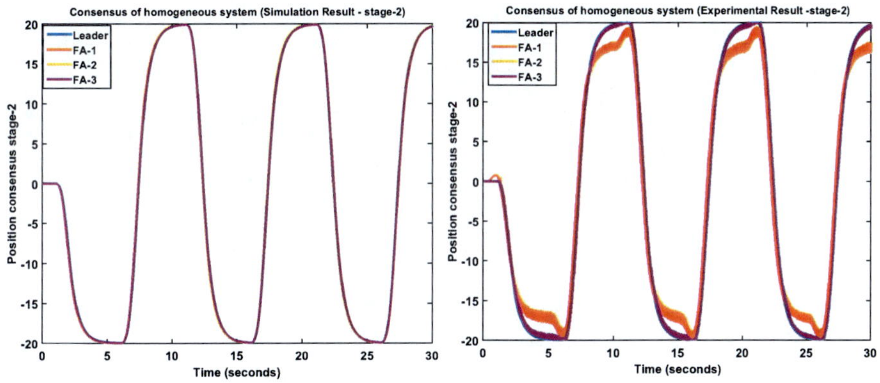

Fig. 5.17 Position consensus (θ_{21}) of 2DOFSFJ robotic arm stage-2

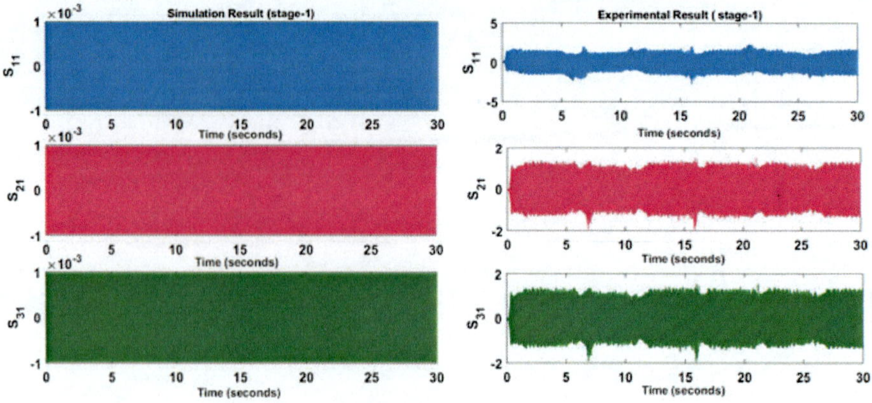

Fig. 5.18 Sliding variable of individual agent of 2DOFSFJ robotic arm stage-1

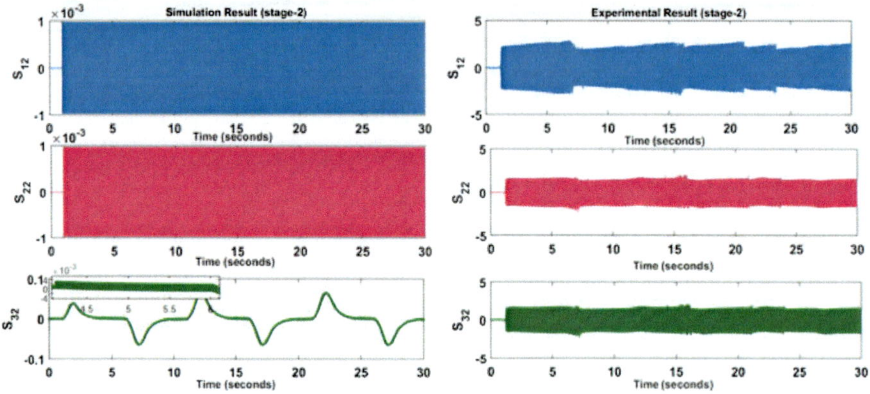

Fig. 5.19 Sliding variable of individual agent of 2DOFSFJ robotic arm stage-2

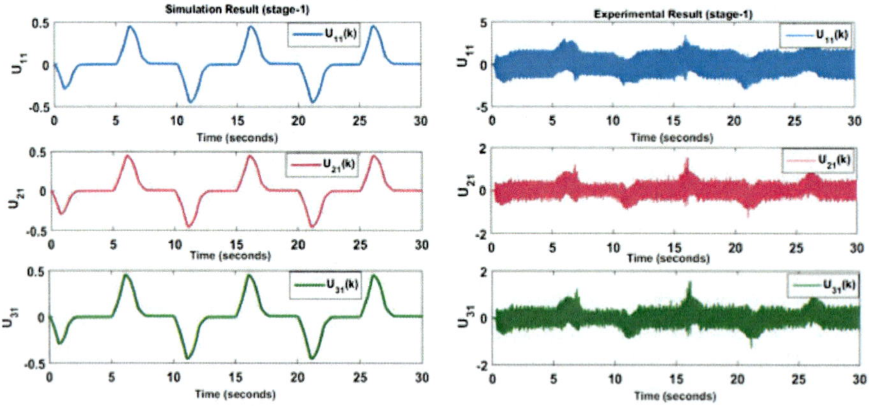

Fig. 5.20 Control effort (u) of 2DOFSFJ robotic arm stage-1

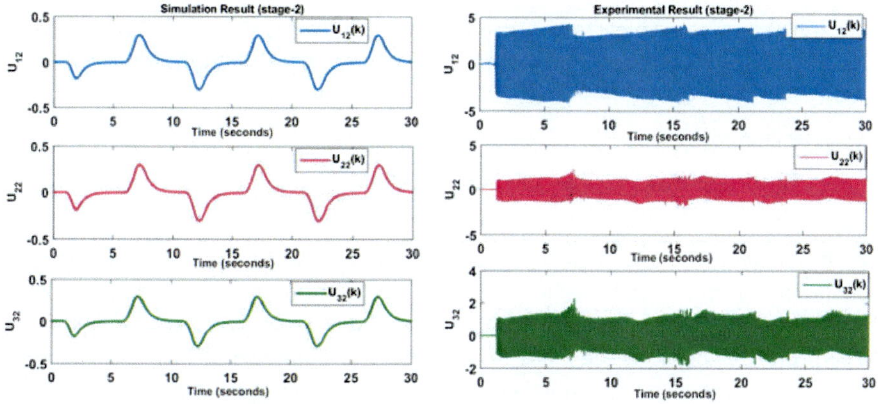

Fig. 5.21 Control effort (u) of 2DOFSFJ robotic arm stage-2

Table 5.2 Error performance index (ISE) for simulation study

	Stage-1			Stage-2		
ISE	e_{11}	e_{21}	e_{31}	e_{12}	e_{22}	e_{32}
Higher order DSM with RL	2.04×10^{-16}	2.04×10^{-16}	2.04×10^{-16}	6.17×10^{-16}	6.17×10^{-16}	6.17×10^{-16}
DSTA	8.80×10^{-16}	2.58×10^{-16}	2.58×10^{-16}	7.80×10^{-12}	7.80×10^{-12}	7.80×10^{-12}
DFSMC	2.84×10^{-12}	2.84×10^{-12}	2.84×10^{-12}	1.25×10^{-13}	1.25×10^{-13}	1.11×10^{-3}.

Table 5.3 Error performance index (ISE) for experimental study

	Stage-1			Stage-2		
ISE	e_{11}	e_{21}	e_{31}	e_{12}	e_{22}	e_{32}
Higher order DSM with RL	2.60×10^{-4}	2.60×10^{-4}	2.60×10^{-4}	2.03×10^{-3}	2.03×10^{-3}	2.03×10^{-3}
DSTA	3.77×10^{-4}	8.34×10^{-4}	8.34×10^{-4}	2.43×10^{-3}	4.14×10^{-3}	4.14×10^{-3}
DFSMC	8.76×10^{-4}	4.46×10^{-4}	4.46×10^{-4}	4.25×10^{-2}	4.48×10^{-2}	4.48×10^{-2}.

posed higher order DSM protocols outperform the DFSM protocol. The detailed performance analysis is presented in Tables 5.2 and 5.3 below:

Tables 5.2 and 5.3 show the simulation and experimental error performance indices for the leader-following consensus of homogeneous DMAS using the proposed two protocols and DFSM protocol. It is inferred from the tables that the protocol having reaching law approach is more efficient compared to other protocols. The results also reveal that DSTA protocol performs better compared to DFSM protocol.

5.5 Conclusion

In this chapter, two topological higher order discrete sliding mode (DSM) protocols are proposed for the leader-following consensus of the homogeneous Discrete Multi-agent System (DMAS) configured with a fixed, directed interaction graph topology. The protocols are designed using (i) reaching law approach and (ii) discrete super-twisting algorithm for the global consensus of homogeneous DMAS. The proposed protocols achieve the consensus of homogeneous DMAS in finite time. The protocols are validated in simulation as well as experimentally on homogeneous DMAS comprising 2-DOF serial flexible joint robotic arms. The comparative results reveal that the proposed higher order DSM protocol using the reaching law approach outperforms the higher order DSM protocol with the super-twisting algorithm. Further, the consensus performance due to higher order DSM protocols is better than consensus due to DFSM protocol.

References

1. Chakrabarty, S., Bandyopadhyay, B., Bartoszewicz, A.: Discrete-time sliding mode control with outputs of relative degree more than one. In: Recent Developments in Sliding Mode Control Theory and Applications, InTech (2017)
2. Salgado, I., Kamal, S., Bandyopadhyay, B., Chairez, I., Fridman, L.: Control of discrete time systems based on recurrent super-twisting-like algorithm. ISA Trans. **64**, 47–55 (2016)
3. Gao, W., Wang, Y., Homaifa, A.: Discrete-time variable structure control systems. IEEE Trans. Indust. Electron. **42**, 117–122 (1995). April
4. Sharma, S., Srivastava, D., Patel, K., Mehta, A.: Design and implementation of second order sliding mode controller for 2-dof flexible robotic link. In: 2018 2nd International Conference on Power, Energy and Environment: Towards Smart Technology (ICEPE), pp. 1–6 (2018)
5. Patel, K., Mehta, A.: Discrete higher order sliding mode protocol for leader-following consensus of heterogeneous discrete multi-agent system. In: Lecture Notes in Electrical Engineering, pp. 1–10. Springer, Singapore (2019)
6. Patel, K., Mehta, A.: Discrete-time event-triggered higher order sliding mode control for consensus of 2-dof robotic arms. Eur. J. Control (2020)

Chapter 6
Event-Triggered Discrete-Time Higher Order Sliding Mode Protocol for Leader-Following Consensus of Homogeneous DMAS

6.1 Event-Triggered Leader-Following Consensus for DMAS

To design an event-triggered higher order DSM protocol using DSTA for leader-following consensus of homogeneous DMAS, such that all the follower agents (3.3) attain the consensus with the leader agent (5.1) by sharing the neighborhood agent information using fixed, directed graph topology [1, 2].

The main objective is to find the control updates using event-triggering condition defined in Eq. (6.3) for each follower agent that tracks the leader trajectory. For this purpose, let us define the local neighborhood state error for the ith agent as

$$\bar{e}_i(k) = \sum_{j=1}^{N} a_{ij}[x_i(k) - x_j(k)] + a_{i0}[x_i(k) - x_0(k)]. \tag{6.1}$$

It may be noted that in the event-triggered control approach, each follower agent updates the control effort at only event-triggering instant $\{k_p^i\}$, $i = 1, 2, ...N$ (which are not periodic) which is obtained using the event-triggering function defined in Eq. (6.3). Then, an event-based local neighborhood state error is defined as

$$\bar{e}_i(k_p^i) = \sum_{j=1}^{N} a_{ij}[x_i(k_p^i) - x_j(k_p^i)] + a_{i0}[x_i(k_p^i) - x_0(k_p^i)], \tag{6.2}$$

where $k \in [k_p^i, k_{p+1}^i)$ and k_p^i is pth the event-triggering scheme for the ith agent is shown in Fig. 6.1.

Now, let us define the event-triggering function for ith agent as

$$k_{p+1}^i = inf\{k : k > k_p^i, g_i(k) \geq 0\}, \tag{6.3}$$

K. Patel and A. Mehta, *Discrete-Time Sliding Mode Protocols for Discrete Multi-Agent System*, Studies in Systems, Decision and Control 303,
https://doi.org/10.1007/978-981-15-6311-9_6

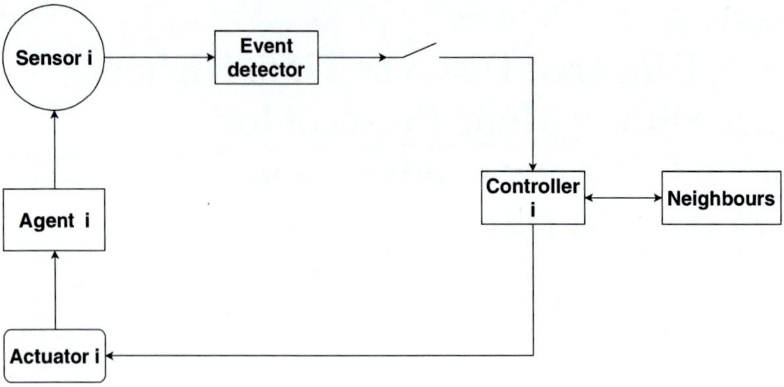

Fig. 6.1 Event-triggered scheme for ith agent

where

$$g_i(k) = ||\hat{E}_i(k)|| - c_1, \; \hat{E}_i(k) = \bar{e}_i(k_p^i) - \bar{e}_i(k), 0 < c_1 < 1$$

in which c_1 is arbitrary constant. From Eq. (6.1) and considering Lemma 3.1, the global consensus error can be written as

$$\bar{\gamma}(k) = ((I_N + \tilde{D} + \tilde{\mathcal{B}})^{-1})(\tilde{L} + \tilde{\mathcal{B}})) \otimes I_n)\tilde{x}_l, \tag{6.4}$$

where $\tilde{x}_l = X(k) - 1_N \otimes x_0(k) \in \mathbb{R}^{nN}$. The eigenvalues of the weighted matrix $\tilde{\xi} = ((I_N + \tilde{D} + \tilde{\mathcal{B}})(\tilde{L} + \tilde{\mathcal{B}}))$ strictly remain inside the unit circle as per the Gershgorin circle criteria [3].

Hence, the global event-triggered-based consensus error is defined as

$$\bar{\gamma}(k_p) = ((I_N + \tilde{D} + \tilde{\mathcal{B}})^{-1})(\tilde{L} + \tilde{\mathcal{B}})) \otimes I_n)\tilde{x}_l, \tag{6.5}$$

where $\tilde{x}_l = X(k_p) - 1_N \otimes x_0(k_p) \in \mathbb{R}^{nN}$ and $k_p = I_{N\times N} \times [k_p^1, k_p^2, k_p^3 ... k_p^N]^{\mathcal{T}}$.

The global consensus error defined in Eq. (6.4) may be written in compressed form as

$$\bar{\gamma}(k) = (\tilde{\xi} \otimes I_n)\tilde{x}_l. \tag{6.6}$$

Considering $(\tilde{\xi} \otimes I_n) = \bar{\upsilon}$ and inserting the value of $\tilde{x}$, we can get

$$\bar{\gamma}(k) = \bar{\upsilon}(X(k) - 1_N \otimes x_0(k)). \tag{6.7}$$

Remark 6.1 The control effort of ith agent is updated only when the event-triggering condition (6.3) is satisfied.

6.2 Event-Triggered Higher Order DSM Protocol Using DSTA for the Global Consensus of Homogeneous DMAS

In this section, event-triggered higher order DSM protocol using DSTA for leader-following consensus of homogeneous DMAS is stated in Theorem 6.1 as follows.

Theorem 6.1 *For the homogeneous DMAS (3.3), if the event-triggering condition defined in (6.3) is satisfied then the global consensus protocol to be updated at the triggering instant* $k_p = [k_p^1, k_p^2, k_p^3, ..., k_p^N]^{\mathcal{T}}$ *is given by*

$$u(k_p) = [-(\sigma_s^{\mathcal{T}} \bar{\upsilon} \bar{F})^{-1}[\sigma_s^{\mathcal{T}} \bar{\upsilon} \bar{E} X(k_p) + \sigma_s^{\mathcal{T}} \bar{\upsilon}(-1_N \otimes \tilde{E} x_0(k_p))$$
$$+\tilde{F} u_0(k_p))]] - \bar{\mathcal{D}}(k_p) - h\sqrt{|\,\tilde{S}(k_p)\,|} sgn(\tilde{S}(k_p)) + w(k_p), \quad (6.8)$$

where $(\sigma_s^{\mathcal{T}} \bar{\upsilon} \bar{F})^{-1}$ *is non-singular.* $\sigma_s^{\mathcal{T}} = I_N \otimes \sigma_{s_i}^{\mathcal{T}}$ *and* $k_p = I_{N\times N} \times [k_p^1, k_p^2, k_p^3 ... k_p^N]^{\mathcal{T}}$..

Remark 6.2 The follower agent is triggered asynchronously based solely on local information, without triggering the leader agent. So, $u_0(k) = u_0(k_p)$ and $x_0(k) = x_0(k_p)$.

Proof Let the sliding surface of ith agent is defined as

$$\tilde{S}_i(k) = \sigma_{s_i}^{\mathcal{T}} \bar{e}_i(k), \quad (6.9)$$

where $\sigma_{s_i}^{\mathcal{T}}$ is a sliding gain that is identified using the LQ method. Now from Eq. (6.9), the global sliding surface can be represented as

$$\tilde{S}(k) = \sigma_s^{\mathcal{T}} \bar{\gamma}(k). \quad (6.10)$$

Using Eq. (6.7), we may write

$$\tilde{S}(k) = \sigma_s^{\mathcal{T}} \bar{\upsilon}(X(k) - 1_N \otimes x_0(k)). \quad (6.11)$$

Further,

$$\tilde{S}(k+1) = \sigma_s^{\mathcal{T}} \bar{\upsilon}(X(k+1) - 1_N \otimes x_0(k+1)). \quad (6.12)$$

Substituting Eqs. (3.3) and (5.1) into Eq. (6.12), we get

$$\tilde{S}(k+1) = \sigma_s^{\mathcal{T}} \bar{\upsilon}(\bar{E} X(k) + \bar{F}(u(k) + \bar{\mathcal{D}}(k)) - 1_N \otimes \tilde{E} x_0(k) + \tilde{F} u_0(k)). \quad (6.13)$$

The proposed protocol $u(k)$ defined in Eq. (6.8) comprises two parts $u_{eq}(k)$ and $u_s(k)$, where $u_{eq}(k)$ is equivalent control and $u_s(k)$ is the super-twisting control. To derive the $u_{eq}(k)$, let us consider $\tilde{s}(k+1) = 0$.

Hence, we may write Eq. (6.13) as

$$0 = \sigma_s^{\mathcal{T}} \bar{\upsilon}(\bar{E}X(k) + \bar{F}(u_{eq}(k) + \bar{\mathcal{D}}(k)) - 1_N \otimes \tilde{E}x_0(k) + \tilde{F}u_0(k). \quad (6.14)$$

And,

$$u_{eq}(k) = [-(\sigma_s^{\mathcal{T}} \bar{\upsilon}\bar{F})^{-1}[\sigma_s^{\mathcal{T}} \bar{\upsilon}\bar{E}X(k) + \sigma_s^{\mathcal{T}} \bar{\upsilon}(-1_N \otimes \tilde{E}x_0(k) + \tilde{F}u_0(k))]] - \bar{\mathcal{D}}(k). \quad (6.15)$$

Now, inspired from [4], we may define a discrete super-twisting algorithm for follower agent as

$$\begin{aligned} u_s(k) &= -h\sqrt{\mid \tilde{S}(k) \mid} sgn(\tilde{S}(k)) + w(k) \\ w(k+1) &= w(k) - \mathcal{F}\tau sgn(\tilde{S}(k)), \end{aligned} \quad (6.16)$$

where $h = [h_1, h_2, h_3, ...h_N] \in \mathbb{R}^{nN}$, $\mathcal{F} = [\mathcal{F}_1, \mathcal{F}_2, \mathcal{F}_3, ...\mathcal{F}_N] \in \mathbb{R}^{nN}$ are gain parameters, and τ is the sampling time. Now using Eqs. (6.15) and (6.16) the control effort required for the global consensus can be written as

$$\begin{aligned} u(k) = [&-(\sigma_s^{\mathcal{T}} \bar{\upsilon}\bar{F})^{-1}[\sigma_s^{\mathcal{T}} \bar{\upsilon}\bar{E}X(k) + \sigma_s^{\mathcal{T}} \bar{\upsilon}(-1_N \otimes \tilde{E}x_0(k) \\ &+\tilde{F}u_0(k))]] - \bar{\mathcal{D}}(k) - h\sqrt{\mid \tilde{S}(k) \mid} sgn(\tilde{S}(k)) + w(k). \end{aligned} \quad (6.17)$$

It may be noted that the protocol (6.17) is updated only at the event-triggering instant k_p and held constant up to next $k_p + 1$ for $\forall\, k \in [k_p, k_{p+1})$ event-triggering instant. However, $k_p + 1$ is not necessarily equal to $k + 1$. Hence, we may write event-triggered global leader-following consensus protocol as

$$\begin{aligned} u(k_p) = [&-(\sigma_s^{\mathcal{T}} \bar{\upsilon}\bar{F})^{-1}[\sigma_s^{\mathcal{T}} \bar{\upsilon}\bar{E}X(k_p) + \sigma_s^{\mathcal{T}} \bar{\upsilon}(-1_N \otimes \tilde{E}x_0(k_p) \\ &+\tilde{F}u_0(k_p))]] - \bar{\mathcal{D}}(k_p) - h\sqrt{\mid \tilde{S}(k_p) \mid} sgn(\tilde{S}(k_p)) + w(k_p). \end{aligned} \quad (6.18)$$

This completes the proof.

Next, the global stability condition for the global consensus of homogeneous DMAS using proposed higher order DSM protocol is derived in the form of Theorem 6.1 as follows.

Theorem 6.2 *The global stability of leader-following consensus of homogeneous DMAS using the proposed event-triggered protocol defined in (6.18) is guaranteed, once the global error dynamics defined in (3.9) drives toward the global sliding* *surface (6.11) and maintains on it for gain* $h, \mathcal{F} > 0$, *provided the following condition is satisfied:*

$$0 \leq \bar{\alpha} < \tilde{S}^{\mathcal{T}}(k_p)\tilde{S}(k_p), \tag{6.19}$$

where $\bar{\alpha} = [-h\sqrt{|\tilde{S}(k_p)|}sgn(\tilde{S}(k_p)) + w(k_p)]^{\mathcal{T}} * -h\sqrt{|\tilde{S}(k_p)|}sgn(\tilde{S}(k_p)) + w(k_p)]$.

Proof Let us define global sliding surface using Eq. (6.11)

$$\tilde{S}(k) = \sigma_s^{\mathcal{T}}\bar{\upsilon}(X(k) - 1_N \otimes x_0(k)). \tag{6.20}$$

Further,

$$\tilde{S}(k+1) = \sigma_s^{\mathcal{T}}\bar{\upsilon}(X(k+1) - 1_N \otimes x_0(k+1))). \tag{6.21}$$

Now selecting the Lyapunov function as

$$\bar{\mathcal{V}}(k) = \tilde{S}^{\mathcal{T}}(k)\tilde{S}(k), \tag{6.22}$$

forward derivative function of (6.22) is defined as

$$\Delta\mathcal{V}_s(k) = \mathcal{V}(k+1) - \mathcal{V}(k), \tag{6.23}$$

$$\Delta\mathcal{V}_s(k) = \tilde{S}^{\mathcal{T}}(k+1)\tilde{S}(k+1) - \tilde{S}^{\mathcal{T}}(k)\tilde{S}(k). \tag{6.24}$$

To ensure the stability, $\Delta\mathcal{V}_s(k) < 0$.

Now substituting $\tilde{S}(k+1)$ in Eq. (6.21) into Eq. (6.24), we may get

$$\begin{aligned}\Delta\mathcal{V}_s(k) = {} & [\sigma_s^{\mathcal{T}}\bar{\upsilon}(X(k+1) - 1_N \otimes x_0(k+1))]^{\mathcal{T}}[\sigma_s^{\mathcal{T}}\bar{\upsilon}(X(k+1) - 1_N \otimes x_0(k+1))] \\ & -\tilde{S}^{\mathcal{T}}(k)\tilde{S}(k).\end{aligned} \tag{6.25}$$

Substituting $X(k+1)$ and $x(k+1)$ using Eqs. (3.3) and (5.1), we get

$$\begin{aligned}\Delta\mathcal{V}_s(k) = {} & [\sigma_s^{\mathcal{T}}\bar{\upsilon}[(\bar{E}X(k) + \bar{F}(u(k) + \bar{\mathcal{D}}(k)) - 1_N \otimes \tilde{E}x_0(k)) + \tilde{F}u_0(k)]]^{\mathcal{T}} \\ & [\sigma_s^{\mathcal{T}}\bar{\upsilon}[(\bar{E}X(k) + \bar{F}(u(k) + \bar{\mathcal{D}}(k)) - 1_N \otimes \tilde{E}x_0(k)) + \tilde{F}u_0(k)]] \\ & -\tilde{S}^{\mathcal{T}}(k)\tilde{S}(k).\end{aligned} \tag{6.26}$$

The protocol is only updated at $k = k_p$ event instant under the event-triggering condition defined in Eq. (6.3). Hence, we may rewritten Eq. (6.26) as

$$\begin{aligned}\Delta\mathcal{V}_s(k_p) = {} & [\sigma_s^{\mathcal{T}}\bar{\upsilon}[(\bar{E}X(k_p) + \bar{F}(u(k_p) + \bar{\mathcal{D}}(k_p)) - 1_N \otimes \tilde{E}x_0(k_p)) + \tilde{F}u_0(k_p)]]^{\mathcal{T}} \\ & [\sigma_s^{\mathcal{T}}\bar{\upsilon}[(\bar{E}X(k_p) + \bar{F}(u(k_p) + \bar{\mathcal{D}}(k_p)) - 1_N \otimes \tilde{E}x_0(k_p)) + \tilde{F}u_0(k_p)]] \\ & -\tilde{S}^{\mathcal{T}}(k_p)\tilde{S}(k_p).\end{aligned} \tag{6.27}$$

Now, substitute the protocol defined in Eq. (6.18) into Eq. (6.27)

$$\Delta\mathcal{V}_s(k_p) = [-h\sqrt{|\tilde{S}(k_p)|}sgn(\tilde{S}(k_p)) + w(k_p)]^{\mathcal{T}} * [-h\sqrt{|\tilde{S}(k_p)|}sgn(\tilde{S}(k_p)) + w(k_p)] - \tilde{S}^{\mathcal{T}}(k_p)\tilde{S}(k_p), \tag{6.28}$$

denoting

$$\bar{\alpha} = [-h\sqrt{|\tilde{S}(k_p)|}sgn(\tilde{S}(k_p)) + w(k_p)]^{\mathcal{T}} * -h\sqrt{|\tilde{S}(k_p)|}sgn(\tilde{S}(k_p)) + w(k_p)].$$

$$\Delta\mathcal{V}_s(k_p) = \bar{\alpha} - \tilde{S}^{\mathcal{T}}(k_p)\tilde{S}(k_p). \tag{6.29}$$

From Eq. (6.29), the term $\bar{\alpha}$ can be set nearer to zero by selecting the proper value of h and $\mathcal{F}$ such that $\Delta\mathcal{V}_s(k) < 0$. Hence, the global stability of homogeneous DMAS is guaranteed.

This completes the proof.

6.3 Design and Implementation of Higher Order DSM Protocol on 2-DOF Serial Flexible Joint Robotic Arm

6.3.1 Experimental Implementation of Higher Order DSM Protocol

For experimental validation of proposed event-triggered protocol, let us consider a leader-following homogeneous DMAS comprising four 2-DOF serial flexible joint robotic arms [5] which is given in Sect. 5.3.1 For validation, one follower agent is an actual 2DOFSFJ robotic arm and other two follower agents and leader are the virtual 2DOFSFJ robotic arm. The experimental setup is shown in Fig. 6.3. In this study, position (θ_{11}) of stage-1 and position (θ_{21}) of stage-2 are to be considered as a single agent parameter for consensus. The simulation and experimental study are carried out using MATLAB R15 interfacing with QUARC software. QUARC is the most suitable software to design, develop, and validate applications in the real-time domain on hardware using MATLAB Simulink. QUARC generates real-time code directly from Simulink-designed controllers and runs it in real time on the windows target all without digital signal processing or without writing a single line of code.

We consider homogeneous DMAS with three follower agent vertices and one leader agent vertex as directed graph topology. The leader agent vertex index is given as 0 while follower agent vertices index are given as 1, 2, and 3, respectively. Their interaction graph topology is shown in Fig. 6.2. The square signals with amplitude of 30 (deg) with frequency of 0.1 Hz for stage-1 of 2DOFSFJ and an amplitude of

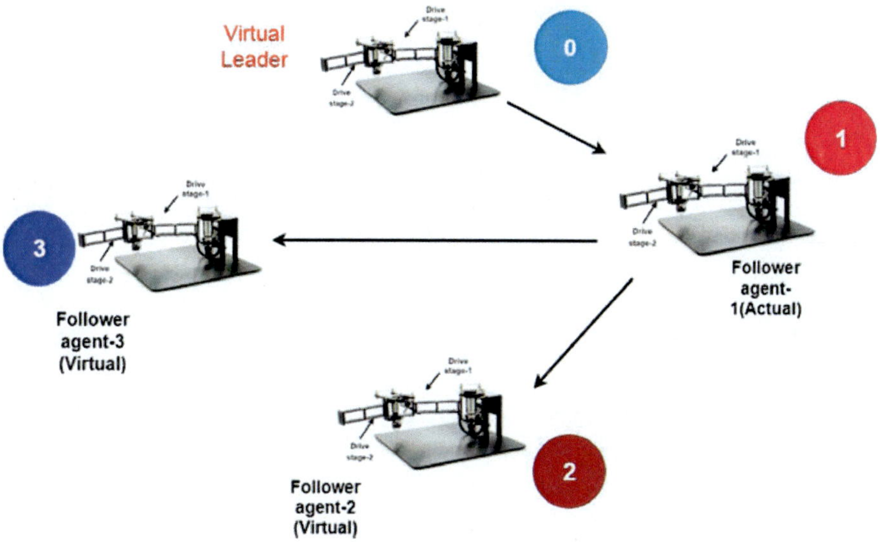

Fig. 6.2 Communication topology

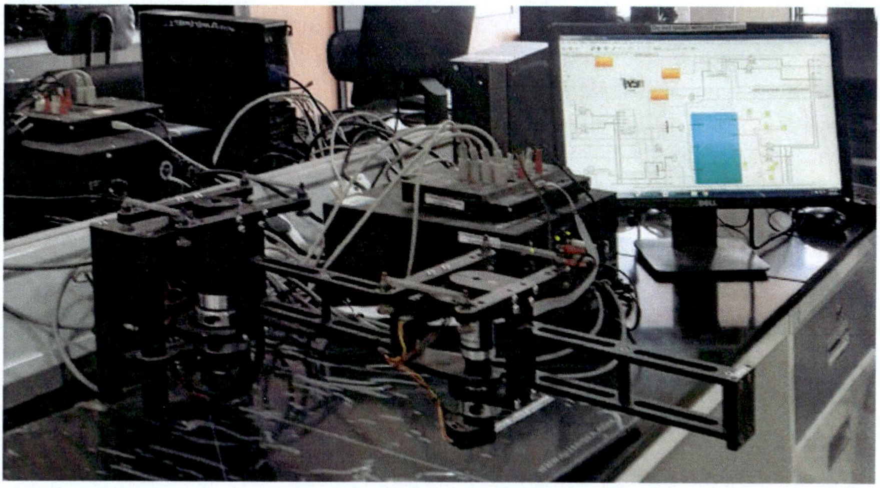

Fig. 6.3 Experimental setup for leader-follower consensus

20 (deg) with frequency of 0.1 Hz for stage-2 are given to the leader agent as a reference. The Laplacian matrix $\tilde{L}$, adjacency matrix $\mathscr{A}$, pinning gain matrix $\tilde{\mathscr{B}}$, and degree matrix $\tilde{D}$ are given as

$$\tilde{L} = \begin{bmatrix} 2 & 0 & 0 \\ -1 & 1 & 0 \\ -1 & 0 & 1 \end{bmatrix}, \mathscr{A} = \begin{bmatrix} 0 & 0 & 0 \\ 1 & 0 & 0 \\ 1 & 0 & 0 \end{bmatrix},$$

$$\tilde{\mathcal{B}} = \begin{bmatrix} 1\,0\,0 \\ 0\,0\,0 \\ 0\,0\,0 \end{bmatrix}, \tilde{\mathcal{D}} = \begin{bmatrix} 2\,0\,0 \\ 0\,1\,0 \\ 0\,0\,1 \end{bmatrix}. \tag{6.30}$$

The robustness property of the derived protocol for leader-following homogeneous DMAS is checked by applying a slowly varying matched disturbance with magnitude $\tilde{\mathcal{D}}_i(k) = 0.0002 * cos(0.01k)$ to ith follower agent.

6.4 Results Discussion

In this section, simulation and experimental results are discussed for higher order DSM protocol using DSTA for the consensus of homogeneous DMAS.

6.4.1 *Simulation and Experimental Results*

Event-triggered higher order DSM protocol defined in (6.18) is applied to the leader-following homogeneous DMAS comprising 2-DOF flexible joint robotic arm. The controller gains for follower agents are chosen as $h = [0.0003\ \ 0.0003\ \ 0.0003]$ and $\mathcal{F} = [0.015\ \ 0.015\ \ 0.015]$, respectively. The sliding gains for surface of ith follower agent for the stage-1 and stage-2 are calculated using LQ method as

$$\sigma_{s_i}^T = \left[23.4923\ -8.3648\ 3.2701\ 2.0258\right],$$

$$\sigma_{s_i}^T = \left[22.0606\ -6.6217\ 1.6634\ 0.4056\right],$$

for stage-1 and stage-2, respectively.

Figures 6.4 and 6.5 show the simulation and experimental results of position consensus for stage-1 and stage-2 using event-triggered higher order DSM protocol with DSTA. It is observed from the result that all the follower agents achieve consensus with the leader agent.

Figures 6.6 and 6.7 show the protocol (u) of each follower agent in the same network for stage-1 and stage-2, respectively, for the consensus using the event-triggering approach and it is clearly observed from the results that the protocol of ith follower agent updated its value if event-triggered condition is satisfied.

Figures 6.8, 6.9, 6.10, 6.11, 6.12 and 6.13 show the inter-event or release time interval for three follower agents for 2DOFSFJ robotic arm stage-1 and stage-2. It is observed from the results that the controller is only updated when the event-triggering condition (6.3) is satisfied. The small window shows the number of events occurred during the two consecutive event-triggering instants of ith follower agent.

Figures 6.14 and 6.15 show the follower agent state error with leader and it is observed that error is bounded.

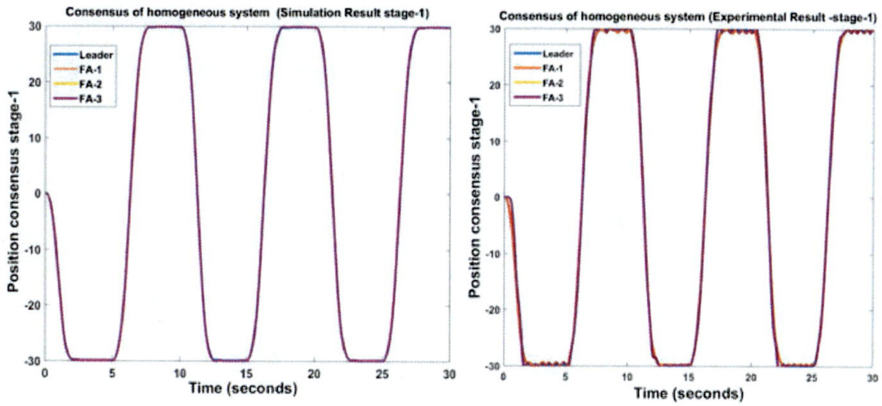

Fig. 6.4 Position consensus (θ_{11}) of 2DOFSFJ robotic arm stage-1

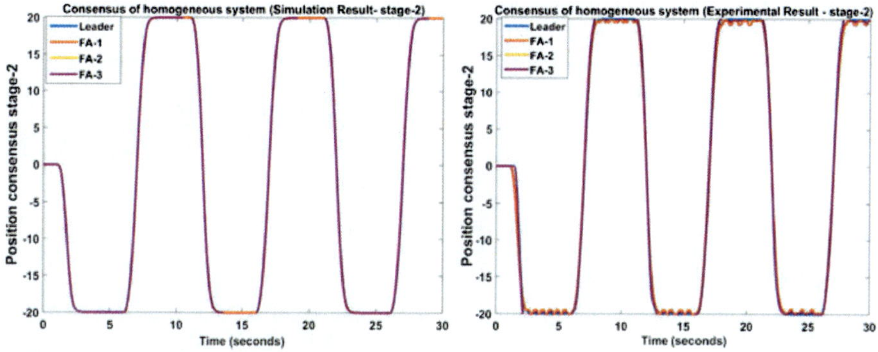

Fig. 6.5 Position consensus (θ_{21}) of 2DOFSFJ robotic arm stage-2

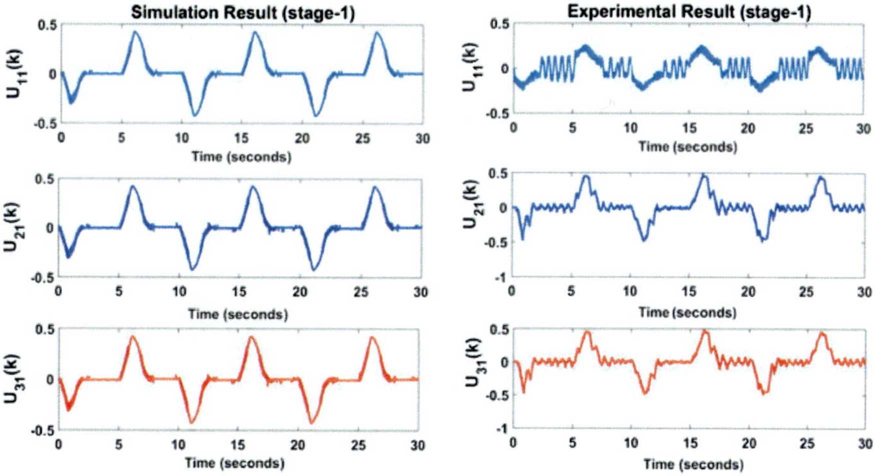

Fig. 6.6 Control effort (u) protocol of 2DOFSFJ robotic arm stage-1

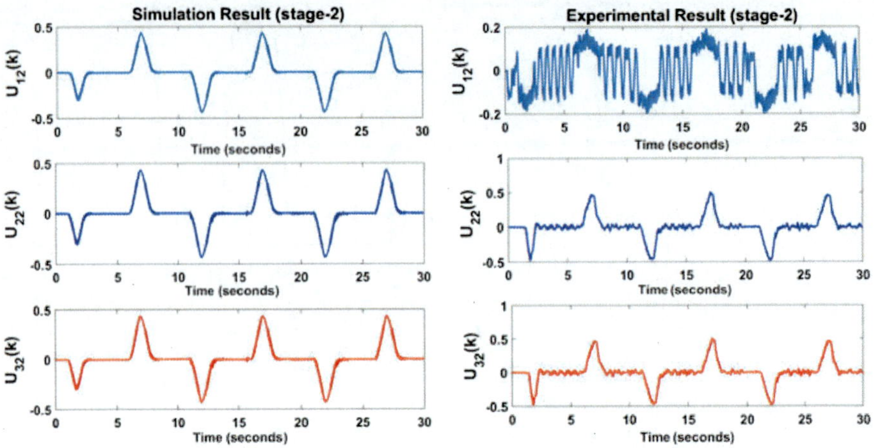

Fig. 6.7 Control effort (u) of 2DOFSFJ robotic arm stage-2

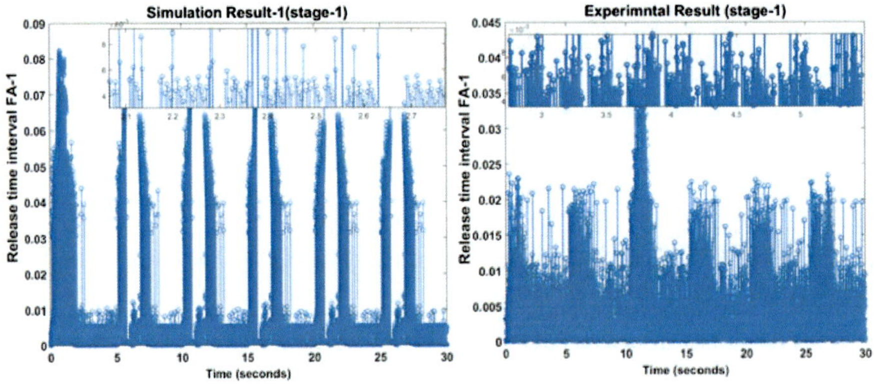

Fig. 6.8 Event-triggering instant of FA-1 for 2DOFSFJ robotic arm stage-1

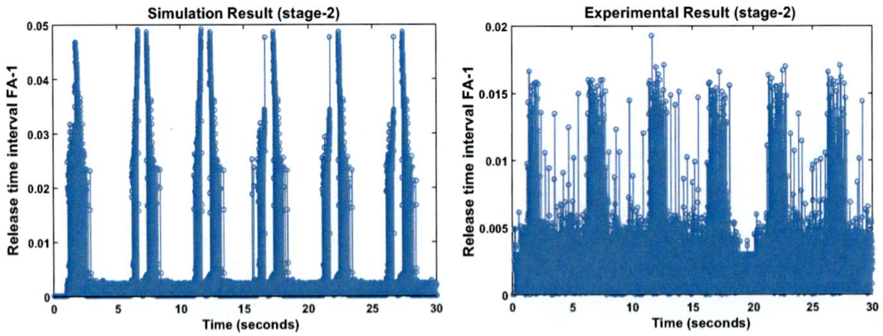

Fig. 6.9 Event-triggering instant of FA-1 2DOFSFJ robotic arm stage-2

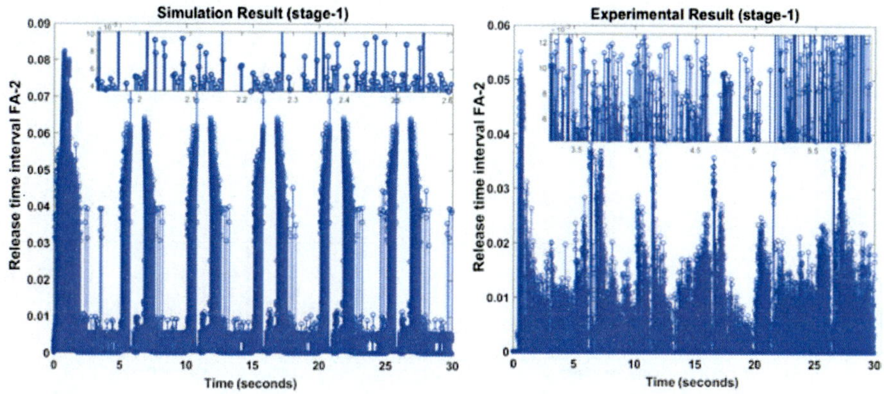

Fig. 6.10 Event-triggering instant of FA-2 for 2DOFSFJ robotic arm stage-1

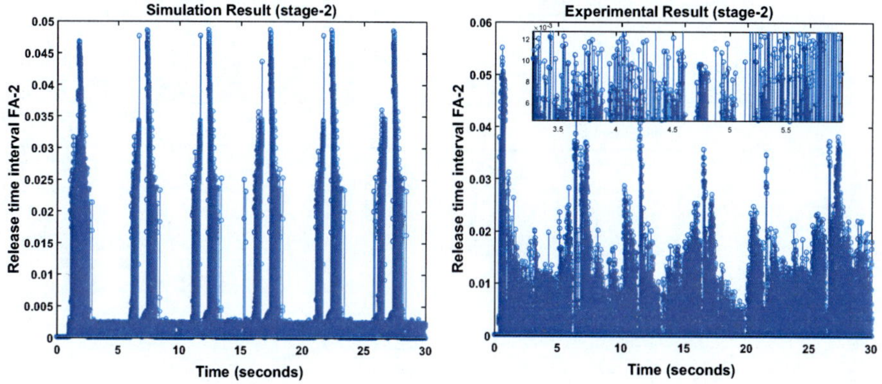

Fig. 6.11 Event-triggering instant of FA-2 for 2DOFSFJ robotic arm stage-2

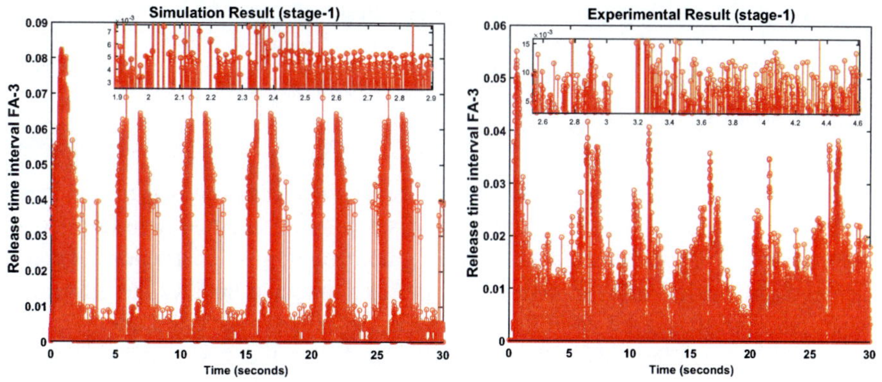

Fig. 6.12 Event-triggering instant of FA-3 for 2DOFSFJ robotic arm stage-1

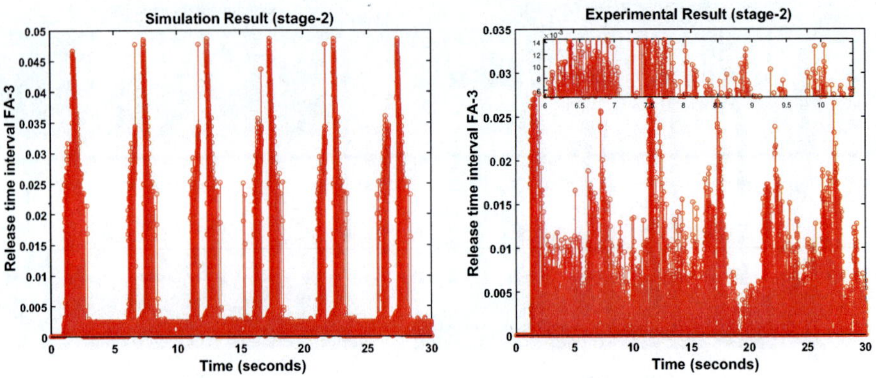

Fig. 6.13 Event-triggering instant of FA-3 for 2DOFSFJ robotic arm stage-2

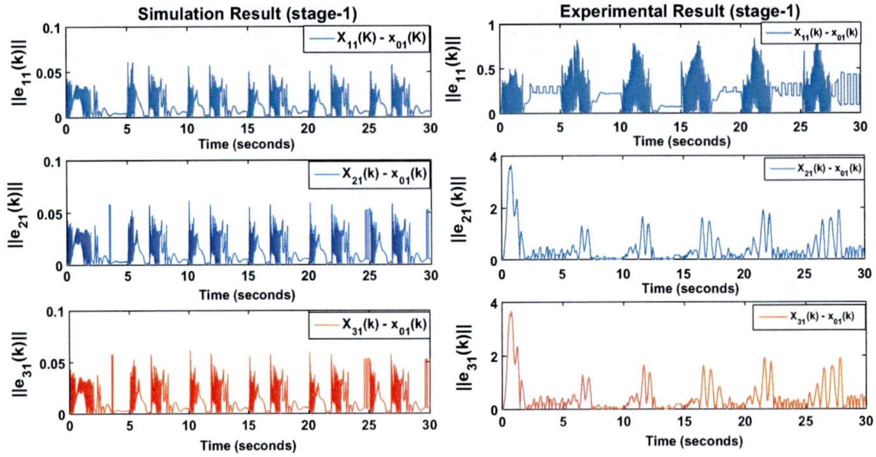

Fig. 6.14 Error between the follower agent and the leader for stage-1

6.5 Conclusion

In this chapter, higher order DSM protocol using DSTA is proposed for the leader-following consensus of the homogeneous discrete multi-agent system DMAS configured with a fixed, directed interaction graph topology. The proposed protocol achieves the leader-following consensus in finite time and updates the information when an event-triggered condition is satisfied in order to save the energy as well as communication bandwidth. Further, the proposed protocol is validated in simulation as well as experimentally on leader-following homogeneous DMAS comprising 2-DOF serial flexible joint robotic arms. The global stability of the proposed event-triggered base protocol is also derived. Finally, the robustness property of protocol is checked by applying slowly varying matched disturbance to the individual follower agents.

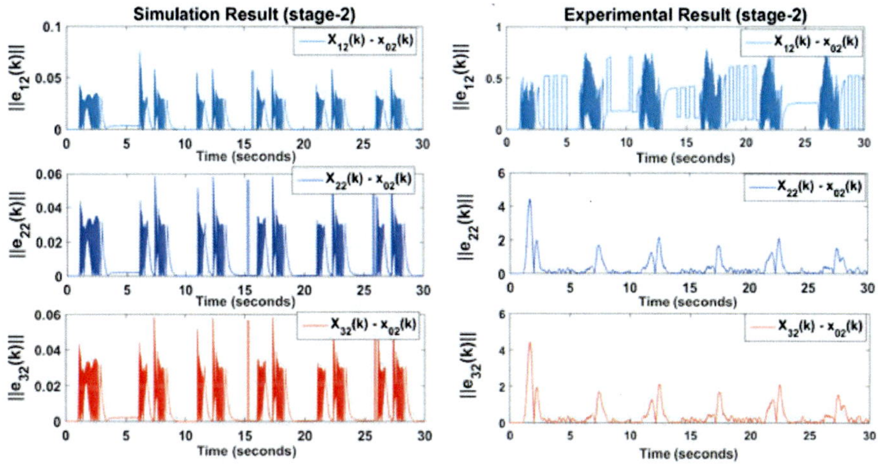

Fig. 6.15 Error between the follower agent and the leader for stage-2

References

1. Patel, K., Mehta, A.: Discrete-time sliding mode control for leader following discrete-time multi-agent system. In: IECON 2018 - 44th Annual Conference of the IEEE Industrial Electronics Society, pp. 2288–2292 (2018)
2. Patel, K., Mehta, A.: Discrete higher order sliding mode protocol for leader-following consensus of heterogeneous discrete multi-agent system. In: Lecture Notes in Electrical Engineering, pp. 1–10. Springer, Singapore (2019)
3. Varga, R.S.: Geršgorin and His Circles. Springer, Berlin (2004)
4. Salgado, I., Kamal, S., Chairez, I., Bandyopadhyay, B., Fridman, L.: Super-twisting-like algorithm in discrete time nonlinear systems. In: The 2011 International Conference on Advanced Mechatronic Systems, pp. 497–502 (2011)
5. Patel, K., Mehta, A.: Discrete-time event-triggered higher order sliding mode control for consensus of 2-dof robotic arms. Eur. J. Control (2020)

Chapter 7
Discrete-Time Higher Order Sliding Mode Protocol for Consensus of Leader-Following Heterogeneous Discrete Multi-Agent System

7.1 Introduction of Heterogeneous Discrete Multi-agent System

Consider the following discrete heterogeneous linear multi-agent system:

$$x_i(k+1) = \tilde{E}x_i(k) + \tilde{F}(u_i(k) + \tilde{\mathcal{D}}_i(k)) \ \forall_i \in N, \tag{7.1}$$

where $i = 1, ..., N$, $\tilde{E} \in \mathbb{R}^{n\times n}$ and $\tilde{F} \in \mathbb{R}^{n\times m}$ are the system matrix and input matrix of ith system, respectively. State vector $x_i(k) \in \mathbb{R}^n$ and the input vector $u_i(k) \in \mathbb{R}^m$, and $\mathcal{D}_i \in \mathbb{R}^m$ is matched disturbance.

Assumption The matrix pair (F, G) for ith system in (7.1) is controllable.

Remark 1 The dynamics of the discrete-time multi-agent system defined in Eq. (7.1) are distinct but their dimensions are identical.

The global heterogeneous DMAS from Eq. (7.1) is defined as

$$X(k+1) = \bar{E}X(k) + \bar{F}(u(k) + \mathcal{D}(k)), \tag{7.2}$$

where $\bar{E} = (I_N \otimes \tilde{E})$, $\bar{F} = (I_N \otimes \tilde{F})$. And $\bar{E} = diag\{E_1, E_2, E_3, ..., E_N\}$, $\bar{F} = diag\{F_1, F_2, F_3, ..., F_N\}$, $X(k) = [x_1(k), x_2(k) ... x_N(k)]^{\mathcal{T}} \in \mathbb{R}^{nN}$ and the input vector $u(k) = [u_1(k), u_2(k) ... u_N(k)]^{\mathcal{T}} \in \mathbb{R}^{mN}$, $\mathcal{D}(k) = [\mathcal{D}_1(k), \mathcal{D}_2(k), \mathcal{D}_3(k), ... \mathcal{D}_N(k)]^{\mathcal{T}} \in \mathbb{R}^{mN}$ matched disturbance vector acting on ith system.

The leader dynamics is defined as

$$x_0(k+1) = \tilde{E}x_0(k) + \tilde{F}u_0(k), \tag{7.3}$$

where $x_0(k) \in \mathbb{R}^n$ is the state vector of the leader.

K. Patel and A. Mehta, *Discrete-Time Sliding Mode Protocols for Discrete Multi-Agent System*, Studies in Systems, Decision and Control 303,
https://doi.org/10.1007/978-981-15-6311-9_7

Problem statement: To design a robust higher order DSM protocol using DSTA for global consensus of heterogeneous DMAS such that all the follower agents (7.2) achieve the consensus with the leader (7.3) using the neighborhood agent information using fixed directed graph topology [1].

The local neighboring error of heterogeneous leader-follower network is defined as

$$\bar{\mu}_i(k) = \sum_{j \in N} a_{ij}[x_i(k) - x_j(k)] + a_{i0}[x_i(k) - x_0(k)]. \tag{7.4}$$

Using Eq. (7.4), the graph-theory-based global consensus error can be rewritten as

$$\bar{\varphi}(k) = ((I_N + \tilde{D} + \tilde{\mathcal{B}})^{-1})(\tilde{L} + \tilde{\mathcal{B}})) \otimes I_n)\tilde{x}_h, \tag{7.5}$$

where $\tilde{x}_h = X(k) - 1_N \otimes x_0(k) \in \mathbb{R}^{nN}$. The eigenvalues of the weighted matrix $\tilde{\xi}_h = ((I_N + \tilde{D} + \tilde{\mathcal{B}})(\tilde{L} + \tilde{\mathcal{B}}))$ are obtained using Gershgorin circle criteria, which is inside the unit circle.

We can express Eq. (7.5) as

$$\bar{\varphi}(k) = (\tilde{\xi}_h \otimes I_n)\tilde{x}. \tag{7.6}$$

Consider $(\tilde{\xi}_h \otimes I_n) = \bar{\Upsilon}_h$ and substituting $\tilde{x}$, we may get

$$\bar{\varphi}(k) = \bar{\Upsilon}_h(X(k) - 1_N \otimes x_0(k)). \tag{7.7}$$

7.2 Higher Order DSM Protocol Using DSTA for the Global Consensus of Heterogeneous DMAS

In this section, a higher order DSM protocol using DSTA for the global heterogeneous DMAS is proposed in the form of Theorem 7.1 as follows.

Theorem 7.1 *For the global heterogeneous DMAS (7.2) is said to achieve the consensus in finite time using higher order DSM protocol with DSTA which is given as*

$$\begin{aligned} u(k) = &-(\bar{\Upsilon}_h\bar{F})^{-1}[\bar{\Upsilon}_h\bar{E}X(k) + \bar{\Upsilon}_h(-1_N \otimes x_0(k+1)) - \tilde{S}(k) \\ &+\bar{h}T\sqrt{|\tilde{S}(k)|}sgn(\tilde{S}(k)) - w(k)] - \bar{\mathcal{D}}(k). \end{aligned} \tag{7.8}$$

Proof The sliding surface for an individual ith agent is defined using Eq. (7.4) as

$$\tilde{S}_i(k) = \bar{\mu}_i(k). \tag{7.9}$$

The global sliding surface is rewritten for the leader-follower network as

$$\tilde{S}(k) = \bar{\varphi}(k). \tag{7.10}$$

Now using Eq. (7.7)

$$\tilde{S}(k+1) = \bar{\Upsilon}_h(X(k+1) - 1_N \otimes x_0(k+1)). \tag{7.11}$$

Further,

$$\tilde{S}(k+1) = \bar{\Upsilon}_h(\bar{E}X(k) + \bar{F}(u(k) + \bar{\mathcal{D}}(k)) - 1_N \otimes x_0(k+1)). \tag{7.12}$$

Let us define the DSTA as [2]

$$\begin{aligned} u_s(k+1) &= -h\sqrt{|\tilde{S}(k)|}sgn(\tilde{S}(k)) + w(k) \\ w(k+1) &= w(k) - \mathcal{F}\tau sgn(\tilde{S}(k)), \end{aligned} \tag{7.13}$$

where $h = [h_1, h_2, h_3, ...h_N] \in \mathbb{R}^{nN}$, $\mathcal{F} = [\mathcal{F}_1, \mathcal{F}_2, \mathcal{F}_3, ...\mathcal{F}_N] \in \mathbb{R}^{nN}$ are gain parameters, and τ is the sampling time.

Then DSTA for global consensus of global heterogeneous DMAS can be defined as

$$\begin{aligned} u(k) = &-[(\bar{\Upsilon}_h\bar{F})^{-1}[\bar{\Upsilon}_h\bar{E}X(k) + \bar{\Upsilon}_h(-1_N \otimes x_0(k+1))]] \\ &- \bar{\mathcal{D}}(k) - h\sqrt{|\tilde{S}(k)|}sgn(\tilde{S}(k)) + w(k). \end{aligned} \tag{7.14}$$

This completes the proof.

7.3 Results Discussion

7.3.1 Simulation and Experimental Results

In this section, we consider total four numbers of agents, among them one follower agent is considered as a 2-DOF serial flexible link robotic arms [3] and other two follower agents and one leader have 2-DOF serial flexible joint robotic arms for the leader-following consensus. In this study, position (θ_{11}) of stage-1 and position (θ_{21}) of stage-2 of respective robotic arms are to be considered as a single agent parameter for the consensus. The simulation and experimental studies are carried out using MATLAB R15 interface with QUARC software The discrete model of 2-DOF serial flexible joint robotic arms for stage-1 and stage-2 is discretized at sampling rate $\tau = 0.002$ and is obtained as

$$\tilde{E}_1 = \begin{bmatrix} 0.9997 & 0.0002696 & 0.001865 & 1.818 \times 10^{-7} \\ 7.81 \times 10^{-5} & 0.9999 & 5.028 \times 10^{-8} & 0.001999 \\ -0.2634 & 0.2634 & 0.868 & 0.0002695 \\ 0.07809 & -0.07809 & 7.455 \times 10^{-5} & 0.9993 \end{bmatrix},$$
$$\tilde{F}_1 = \begin{bmatrix} 0.0002673 \; 3.545 \times 10^{-9} \; 0.2612 \; 7.042e \times 10^{-6} \end{bmatrix},$$
$$\tilde{G}_1 = \begin{bmatrix} 1 \; 0 \; 0 \; 0 \end{bmatrix}. \tag{7.15}$$

Similarly, for stage-2

$$\tilde{E}_2 = \begin{bmatrix} 0.9979 & 0.002078 & 0.001739 & 1.416 \times 10^{-6} \\ 0.0007443 & 0.9993 & 4.63 \times 10^{-7} & 0.001994 \\ -1.983 & 1.983 & 0.7499 & 0.002075 \\ 0.7433 & -0.7433 & 0.0006783 & 0.994 \end{bmatrix},$$
$$\tilde{F}_2 = \begin{bmatrix} 0.0004521 \; 5.827 \times 10^{-8} \; 0.4315 \; 0.0001149 \end{bmatrix},$$
$$\tilde{G}_2 = \begin{bmatrix} 1 \; 0 \; 0 \; 0 \end{bmatrix}. \tag{7.16}$$

Similarly, the state space model for 2-DOF serial flexible link robotic arms for stage-1 and stage-2 is obtained as

$$\tilde{E}_1 = \begin{bmatrix} 1 & 0.001206 & 0.001879 & 8.126 \times 10^{-5} \\ 0 & 0.9983 & 0.0001208 & 0.001999 \\ 0 & 1.181 & 0.8817 & 0.001206 \\ 0 & -1.65 & 0.1182 & 0.9983 \end{bmatrix},$$
$$\tilde{F}_1 = \begin{bmatrix} 0.0002695 \; -0.0002694 \; 0.2639 \; -0.2638 \end{bmatrix},$$
$$\tilde{G}_1 = \begin{bmatrix} 1 \; 0 \; 0 \; 0 \end{bmatrix}. \tag{7.17}$$

$$\tilde{E}_2 = \begin{bmatrix} 1 & 0.003261 & 0.00128 & 4.306 \times 10^{-5} \\ 0 & 0.9947 & 0.0007132 & 0.00193 \\ 0 & 2.763 & 0.3857 & 0.03775 \\ 0 & -4.818 & 0.6043 & 0.9345 \end{bmatrix},$$
$$\tilde{F}_2 = \begin{bmatrix} 0.0004178 \; -0.0004137 \; 0.3563 \; -0.3505 \end{bmatrix},$$
$$\tilde{G}_2 = \begin{bmatrix} 1 \; 0 \; 0 \; 0 \end{bmatrix}. \tag{7.18}$$

Let us consider three heterogeneous follower agents and one leader to represent the communication graph shown in Fig. 7.1. The leader agent node notation is given as 0 while follower agent node notations are given as 1, 2, and 3, respectively. A square signal with amplitude of 20 (deg) and frequency of 0.1 Hz is given to the leader agent for both stages of robotic arms. The diagonal matrix $\tilde{D}$, adjacency matrix $\mathcal{A}$, pinning gain matrix $\tilde{\mathcal{B}}$, and Laplacian matrix $\tilde{L}$ are defined as

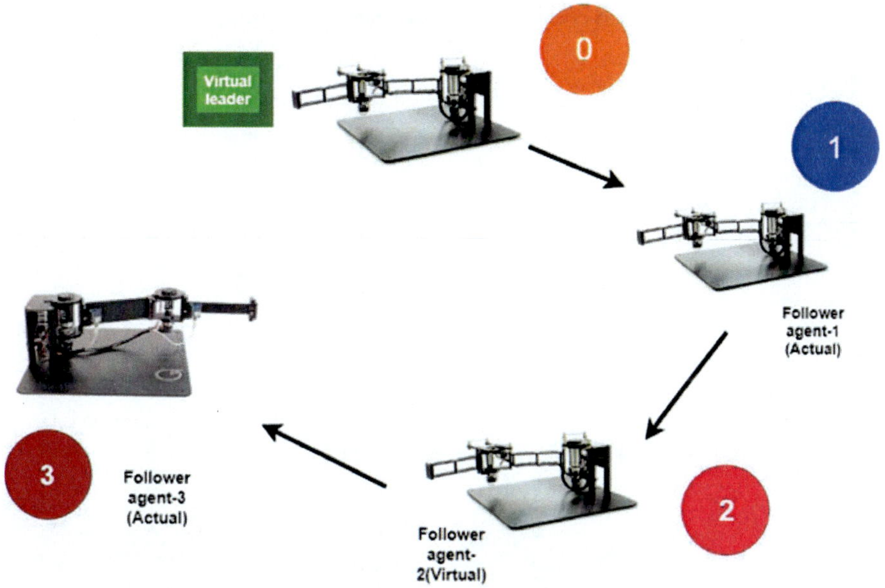

Fig. 7.1 Communication topology

$$\tilde{D} = \begin{bmatrix} 1 & 0 & 0 \\ 0 & 2 & 0 \\ 0 & 0 & 1 \end{bmatrix}, \mathcal{A} = \begin{bmatrix} 0 & 0 & 0 \\ 1 & 0 & 0 \\ 0 & 1 & 0 \end{bmatrix}, \tilde{\mathcal{B}} = diag\{1, 0, 0\}, \tilde{L} = \begin{bmatrix} 1 & 0 & 0 \\ -1 & 2 & 0 \\ 0 & -1 & 1 \end{bmatrix}. \quad (7.19)$$

In order to check the robustness of the derived protocol for global heterogeneous DMAS, a slow varying disturbance is applied to the each DMAS with magnitude $\tilde{\mathcal{D}}_i(k) = 0.002 * cos(0.01k)$ to the follower agent. Gains for the follower agents are chosen as $h = [0.15 \;\; 0.15 \;\; 0.15]$, $\mathcal{F} = [0.0002 \;\; 0.0002 \;\; 0.0002]$ for DSTA.

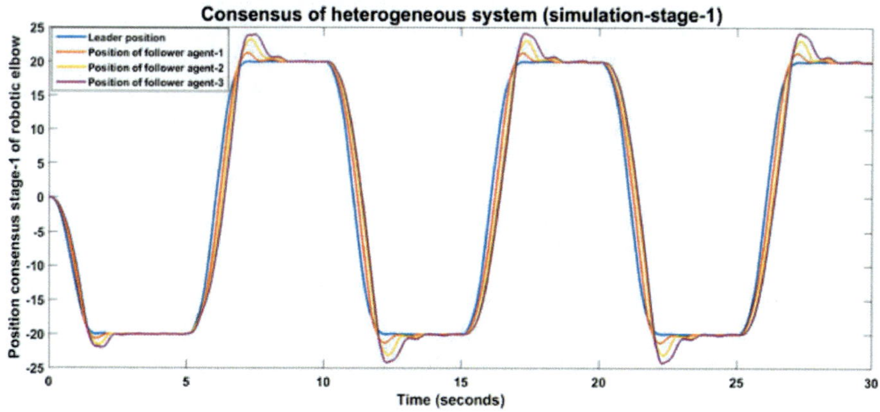

Fig. 7.2 Position (θ_{11}) consensus of 2-DOF robotic system stage-1

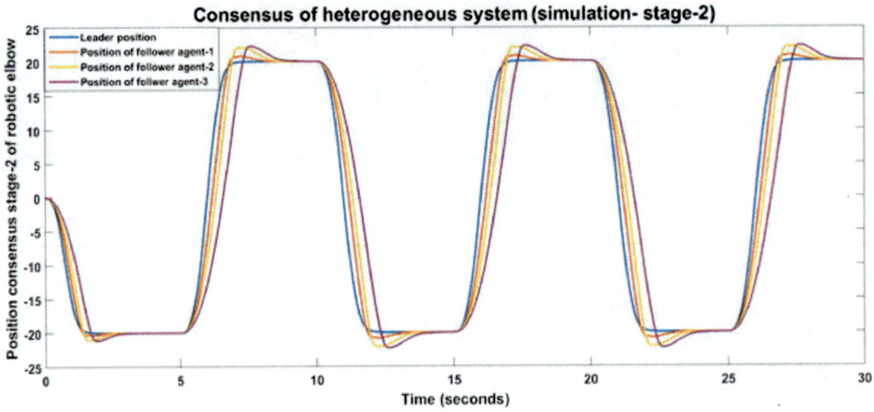

Fig. 7.3 Position (θ_{21}) consensus of 2-DOF robotic system stage-2

Fig. 7.4 Experimental setup for leader-follower consensus

The proposed protocol (7.14) is applied to the leader-following heterogeneous DMAS given in (7.2) for the consensus. Figure 7.2 shows the simulation result of position consensus of 2-DOF robotic heterogeneous system stage-1 using DSTA and finds that all the follower agents synchronize with leader in finite time.

Similarly, Fig. 7.3 shows the simulation result of position consensus of 2-DOF robotic heterogeneous system stage-2.

Figure 7.4 shows the experimental setup available in our laboratory.

Figures 7.5 and 7.6 show experimental validation of the position consensus of the 2-DOF robotic heterogeneous system for stage-1 and stage-2 using DSTA protocol and find that all the follower agents synchronize with the leader in finite time.

Figures 7.7 and 7.8 show the simulation and experimental results of the consensus error.

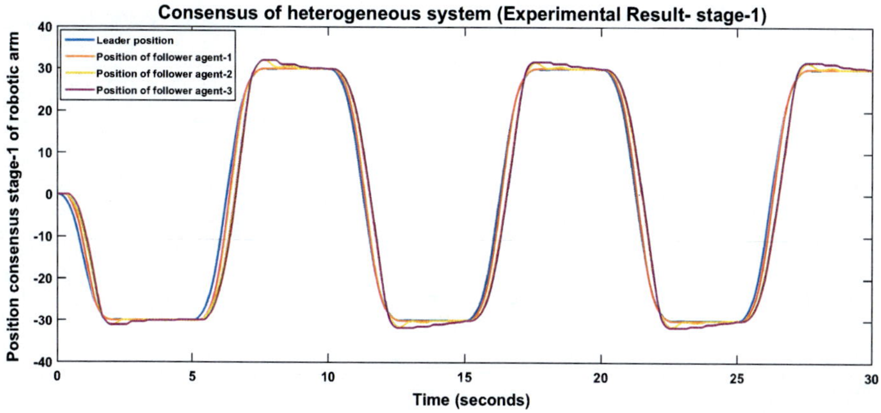

Fig. 7.5 Position (θ_{11}) consensus of 2-DOF robotic system stage-1

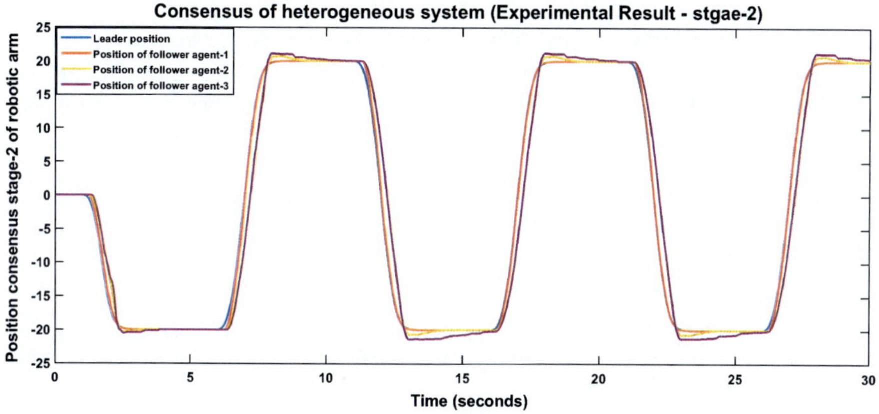

Fig. 7.6 Position (θ_{21}) consensus of 2-DOF robotic system stage-2

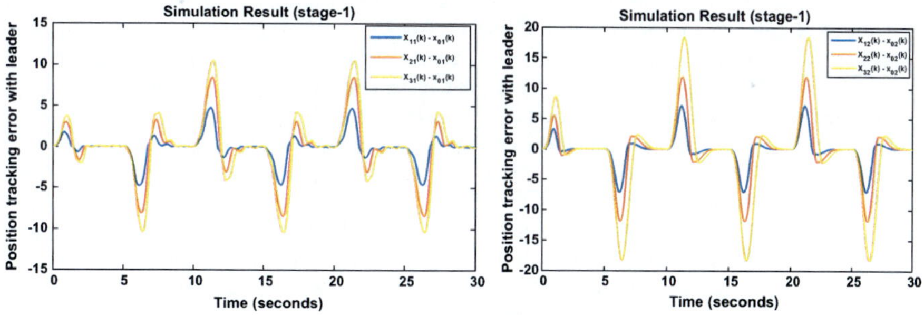

Fig. 7.7 Simulation result of error between the position of the follower agent and the leader

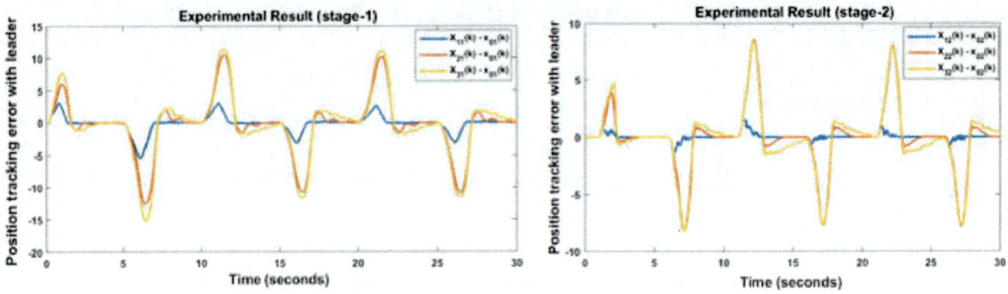

Fig. 7.8 Experimental result of error between the position of the follower agent and the leader

7.4 Conclusion

In this chapter, higher order DSM protocol using DSTA is proposed for leader-following consensus of a discrete heterogeneous multi-agent system configured with a fixed, directed interaction graph topology. The proposed protocol achieves the leader-following consensus of heterogeneous DMAS in finite time. The proposed protocol is validated in simulation as well as experimentally using 2-DOF serial flexible joint and 2-DOF flexible link robotic arms for the consensus.

References

1. Patel, K., Mehta, A.: Discrete-time sliding mode control for leader following discrete-time multi-agent system. In: IECON 2018 - 44th Annual Conference of the IEEE Industrial Electronics Society, pp. 2288–2292 (2018)
2. Patel, K., Mehta, A.: Discrete higher order sliding mode protocol for leader-following consensus of heterogeneous discrete multi-agent system. In: Lecture Notes in Electrical Engineering, pp. 1–10. Springer Singapore (2019)
3. Patel, K., Mehta, A.: Discrete-time event-triggered higher order sliding mode control for consensus of 2-dof robotic arms. Eur. J. Control (2020)
4. Zhang, D., Xu, Z., Karimi, H.R., Wang, Q., Yu, L.: Distributed h_∞ output-feedback control for consensus of heterogeneous linear multiagent systems with aperiodic sampled-data communications. IEEE Trans. Indust. Electron. **65**(5), 4145–4155 (2018)
5. Wang, D., Yu, M.: Leader-following consensus for heterogeneous multi-agent systems with bounded communication delays. In: 2016 14th International Conference on Control, Automation, Robotics and Vision (ICARCV), pp. 1–6 (2016)
6. da Silva, M.M., de Camargo, R.F., Pinheiro, J.R., Vieira, R., Kenne, T.G., Botteron, F.: Discrete-time super-twisting sliding-mode control applied to a dstatcom-based voltage regulator for a self excited induction generator. In: IECON 2014 - 40th Annual Conference of the IEEE Industrial Electronics Society, pp. 4273–4278 (2014)

Chapter 8
Concluding Remarks and Future Scope

This monograph presents novel discrete-time sliding mode protocols for the leader-following consensus of a discrete multi-agent system in the presence of matched disturbances. The DSM protocols are developed for two types of communication graph topologies, namely, (i) fixed graph topology and (ii) switching graph topology.

To begin, two distinct discrete-time sliding mode global topological protocols based on Gao's reaching law and Power rate reaching law are proposed for leader-following consensus of a discrete multi-agent system configured with a fixed, undirected graph topology as a global system having one leader and other agents as followers. Both protocols ensure finite-time consensus of follower agents with the leader. The necessary conditions for the global stability of consensus are derived and the robustness property of the proposed controller is also checked by applying a slowly varying disturbance to DMAS.

Further, the analysis for the number of time steps required for the global consensus of DMAS with both protocols is derived. It is inferred that the time required for the global consensus due to the protocol using Gao's reaching law takes more time in comparison to the protocol using the Power rate reaching law. Also the simulation results reveal that the chattering magnitude is $O(\tau^2)$ in case of Power rate reaching law compared to Gao's reaching law which has chattering magnitude of $O(\tau)$. The protocol using Power rate reaching law improves the consensus performance of DMAS by reducing the chattering effect and increases the speed of consensus of DMAS with a compromise of the invariance property. Both global topological protocols are also implemented in the Simulink environment for the leader-following consensus of a homogeneous multi-agent system comprising 2-DOF (degree of freedom) helicopter systems.

The above discrete-time sliding mode protocols for consensus of a leader-following discrete multi-agent system are extended for the switching graph topologies. The discrete multi-agent system is described with an undirected switching graph

K. Patel and A. Mehta, *Discrete-Time Sliding Mode Protocols for Discrete Multi-Agent System*, Studies in Systems, Decision and Control 303,
https://doi.org/10.1007/978-981-15-6311-9_8

topology as a global system having one leader and other as follower agents. Sliding surface for global consensus of agents for switching graph topology is defined and discrete-time sliding mode protocols using modified Gao's and Power rate reaching laws are derived. Under the influence of switching topology, the graph topology switches in a different number of step intervals and ensures that protocols achieve the consensus in fixed time steps. The condition for global stability in both cases is also derived using the Lyapunov function. Further, the efficacy of the protocols is checked on the multi-agent system comprising multiple 2-DOF (degree of freedom) helicopter systems. Whereas the pitch angle and its velocity and yaw angle and its velocity are single agent parameters used for the consensus. Finally, the robustness property of both algorithms is checked by applying the matched disturbances in follower agents.

It is inferred that both DSM protocols proposed as above possess chattering which deteriorates the consensus performance of homogeneous DMAS. To overcome this problem, two topological higher order Discrete Sliding Mode (DSM) protocols are proposed for the leader-following consensus of the homogeneous Discrete Multi-Agent System (DMAS) configured with a fixed, directed interaction graph topology. The protocols are designed using (i) reaching law approach and (ii) discrete super-twisting algorithm for the global consensus of homogeneous DMAS. The proposed protocols achieve the consensus of homogeneous DMAS in finite time. Both protocols validated in simulation as well as experimentally on homogeneous DMAS comprise of 2-DOF serial flexible joint robotic arms. The comparative results reveal that the proposed higher order DSM protocol using the reaching law approach outperforms the higher order DSM protocol with the super-twisting algorithm. Further, the consensus performance due to higher order DSM protocols is better than consensus due to previously proposed first-order DSM protocols. The robustness property of each protocol is checked by applying slowly varying matched disturbance to the individual follower agents.

It is observed that in many applications the consensus is required only when a certain event occurs. Hence, an event-triggered-based higher order discrete super-twisting sliding mode protocol for the leader-following consensus of DMAS is proposed. The proposed protocol achieves the leader-following consensus in finite time and updates the information when an event-triggered condition is satisfied in order to save the energy as well as communication bandwidth. The proposed protocol is validated in simulation as well as experimentally on leader-following homogeneous DMAS comprising 2-DOF serial flexible joint robotic arms. The global stability of the proposed event-triggered protocol is also derived. In this case, also, the robustness property of protocol is checked by applying slowly varying matched disturbance to the individual follower agents.

All the above protocols are developed for the homogeneous DMAS in which the dynamics of all the system in a MAS configuration are identical. However, in certain applications, the dynamics of the agents may not be identical. To deal with this heterogeneous DMAS, higher order discrete sliding mode control protocol using a discrete-time super-twisting algorithm for the leader-following consensus of heterogeneous DMAS is proposed. The proposed protocol ensures the consensus of heterogeneous follower agents with a leader agent in finite-time step. The proposed protocol is validated in simulation as well as experimentally on heterogeneous DMAS comprising 2-DOF flexible joint and 2-DOF flexible link robotic arms. It is inferred from the simulation and experimental results that the proposed protocol is globally applied to the actual application for the consensus of heterogeneous DMAS. Finally, the robustness property is also checked by applying slowly varying disturbance to the individual follower agents.

- **Future Scope**

In last one decade, a lot of research work has been carried out in the domain of cooperative control. But still there are many challenges need to be addressed especially in the discrete-time domain. Some of the examples are as follows:

- In multi-agent system cooperation, coordination and communication are the most important factors to achieve a common task. However, due to complex environmental conditions such as temperature, aging of equipment due to poor maintenance, and power loss are unavoidable and as a result of that machine crashes, sensor, and actuator failures which will lead the collapse of the entire network of DMAS. With due consideration of multiple adverse effects of failures, one can design consensus protocol using a fault-tolerant control algorithm for the heterogeneous and homogeneous DMAS and enhance the performance and reliability of the entire DMAS. It is also important to study heterogeneous DMAS with distinct dynamics with nonidentical dimensions and the protocol design of such DMAS is quite challenging.
- To achieve coordinated control and consensus of MAS, agents share their information using communication networks. These networks carry a large number of information sources and due to limited network bandwidth and data traffic, all the agents may not receive complete information from their respective neighbor agents. Hence, inaccurate information shared among agents may lead to disrupt the entire consensus performance of DMAS. Therefore, it is important to design a protocol with communication constraints such as time delays, packet dropouts for the consensus of homogeneous or heterogeneous DMAS.
- It is interesting to study the consensus of MAS if some of the agents in the group disagree called bipartite consensus. In such situation, a special type of graph called a signed Laplacian matrix can be conceptualized for the overall communication of MAS. There is need to design a robust DSM protocol for heterogeneous and homogeneous DMAS for bipartite consensus.
- To carry out cooperative missions in realistic environments, the prevention of obstacles/collisions is of practical significance. Thus, in this circumstance, it is important to design a consensus protocol that takes care of the collision of MAS and provides safe consensus among the agents.